THE UNORTHODOX
CREATOR

THE UNORTHODOX CREATOR

CREATOR

HOW TO SURVIVE AND THRIVE IN THE DIGITAL WORLD

DERRON PAYNE

NEW DEGREE PRESS

THE UNORTHODOX CREATOR
How to Survive and Thrive in the Digital World

ISBN   979-8-88504-072-3 *Paperback*
         979-8-88504-700-5 *Kindle Ebook*
         979-8-88504-179-9 *Ebook*

For my late great-grandmother, Esidora Morris. This book is written in her honor.

4/4/1929–2/25/2021

CONTENTS

INTRODUCTION

Picture this: We're ten years into the future, and the world as you know it has changed quite a bit. The little fifteen-year-old boy that lived down the street from you is now one of the most famous and influential people in the world. What is he famous for you might ask? He is the top player in a new game that came out the year before. He streams his gameplay online, and who would've thought he also has a funny side. He has over ten million followers and averages over five hundred thousand viewers while live on a nightly basis. At this point, he is getting more viewers than any cable television station. It makes sense, though, because in this time, seeing someone with a cable box in their home is as rare as seeing a person today with a Walkman or CD player.

Not only has he taken over the internet world, but he also now has his own cereal, which over the past month became the most-sold cereal in the United States. Why? Simply because he has developed such a strong community of supporters that love his content.

That's not all though; his older brother who is about twenty-one years old—and also happens to help manage his brother and some of his investments—just added a new property to their portfolio for a whopping $2.5 million. How many square feet is the place you ask? The answer: Zero. The property is a piece of land in the metaverse, basically the virtual world (this will be explained in greater detail later on in the book).

This dream may seem farfetched, but the world is heading in this direction. We already have influencers breaking into new spaces of business and all because of their internet followings.

Growing up, I never expected my life would be without cable TV. As a kid, television was one of the main places where I would get my entertainment. When I wasn't doing schoolwork, my parents kept me active with extracurricular activities such as tae kwon do, which I did until I was about seven, and basketball, which I still play today for fun. After my responsibilities were taken care of, whatever other time I had I spent playing video games, watching cable TV, and occasionally building things.

Cable TV originated in 1948 but it wasn't until the nineties that it became popular, and it was a staple in my life growing up (California Cable & Telecommunications Association, 2022). I loved cartoons; *Scooby-Doo* was my favorite (Barbera and Hanna et al., 2004).

I remember being a kid and being scared of growing up because I loved cartoons, and I was scared of the crime shows that my parents enjoyed. I thought that when you got older, you had to give up cartoons for more adult shows. I don't remember what age I was when I realized that I actually enjoyed watching those crime shows with my parents and that they were just a more complex version of *Scooby-Doo*. I always enjoyed watching sports with my dad though, specifically basketball. Back then, to watch these types of shows, you needed some form of cable. Years went by, and we started seeing some major changes.

Technology and the internet improved, and as it evolved, new creations were born: streaming platforms and social media platforms, which appeared in the nineties and early 2000s. These platforms gave us access to more content to consume all on our own time. Previously, I would be limited to what was scheduled to play on whatever television channel, but with these new inventions, I had a choice. I got my first laptop when I graduated from middle school, and after a few years, I started consuming more and more internet content.

When it came time for me to graduate, move out, and start making my adult decisions and purchases, I thought long and hard about what choices to make. After getting an apartment, it was time to fill it with the essentials, furniture, and miscellaneous items. Then I made sure to purchase Wi-Fi, but the biggest decision was what I decided not to buy: cable. A lot of my peers have moved out on their own and made the same decision I did because we tend to seek our entertainment from the internet.

It's not just my peers, however, or just the people in my age group; a large number of people are cutting the traditional cable cord. By the end of 2019, a total of 39.3 million people were cord cutters, and that number is supposed to grow to 55.1 million by 2022. Cable TV penetration is at around 81 percent, and within the next decade, that number will fall to around 55 percent (Nick G., 2021).

YouTube is a video-sharing platform where anyone can view and post videos. It has evolved into quite the tool, with many people jumping on the site and creating careers for themselves. What would start as an online video dating site in 2005 would later grow into something that I'm sure the original founders could not imagine (Nieva, 2016). It is now the second-most-visited website, and over five hundred hours of video content were added to its server every minute in 2019 (McFadden, 2021). Content creators on YouTube—better known as YouTubers—created a lane for themselves and continued to improve over the years, to the point that a lot of the videos of the bigger creators now have huge budgets as well as brand deals. The production has been taken to a new level. Creators are making videos giving away houses and cars, doing reviews of other content, producing music, creating short movies and documentaries, and so much more. There is really no limit to what can be produced (provided it's appropriate) and what heights the creators can reach (Hosch, 2021).

Also there is a newer app that launched in 2016 in China as Douyin and later in 2017 was released internationally as TikTok. It is one of the fastest-growing apps ever with over two billion worldwide downloads. It's an app in which people create fifteen-second to ten-minute videos. It may seem a tad less refined than YouTube where videos are longer and creators focus more on editing, but that is exactly why TikTok is doing so well and its creators—better known as TikTokers—are achieving so much success right now. We have come to a time where there is a huge need for instant gratification, so this short-form content is winning, so much so that some of these TikTokers have amassed followings of millions of people. Young people are finding a lot of success in this space; for example, the largest TikToker, Charli D'Amelio, grew a following of over one hundred million people before she was even seventeen years old (Tidy and Smith Galer, 2022).

Twitch started in 2011 as a livestreaming platform primarily for video game players. With the growing interest in video games, interest also grew for watching people play video games. At first, I compared it to watching professional sports. I believe people are infatuated with seeing greatness and watching others perform amazing feats, and I saw no difference with video games. I love video games and have been playing them for years. And even though I couldn't dream of doing some of the amazing things that pros can do, I love to see it. At some point this changed; people still watch to see greatness but now also watch just for entertainment. Some video game streamers aren't very good at video games, but they are very funny and produce laughs. Others stream but don't play games at all because they realized it is all about the content. If you can keep people entertained, they will watch, whether you're making them laugh or doing something crazy (Geeter, 2019).

These platforms grew to be more successful with the changing times and created a new generation of consumers that are different from those that came before them. This generation that grew up having the benefits of new technology that was continuously evolving is so used to instant gratification and typically loses interest with long waits.

While there was once a time when cable television was the main form of home entertainment, that is no longer the case. The content we consumed was decided for us by a major production company of some sort, but today we can choose. The creators of content have become amazing. They are bigger than just some funny actors or actresses in a video; they have become a business. We aren't even at the peak of what the internet will become yet. But based on what has happened, we can make educated guesses on what the future holds, and that is why I wanted to write this book. I am intrigued by this digital world and the opportunities that will come with it.

I am qualified to write this because of the time period in which I grew up; I am in the middle of two completely different generations. I have experienced life with and without all these amazing devices. When I was a kid, the internet was not what it is today. I remember having a gigantic prehistoric desktop as the family computer. I only remember using it maybe once or twice, and it wasn't to use the internet but to play games like minesweeper that typically came predownloaded on the computer. The internet back then was vastly different compared to what it is today; in order to get on the internet back then, we had to use dial, which meant we couldn't get calls while online. The internet was also so much slower than the high-speed internet of today.

As I grew up, so did technology, and having had the luxury of growing with the technology, I learned a lot and also had to adapt to the times. Being put in the position that forced me to learn to adapt is one of the reasons why I am the perfect person for sharing this message.

New-age content creators are some of the biggest marketers and entrepreneurs of our time, and it just so happens that when I went to college, I chose to study marketing and management with a minor in entrepreneurship. I attended the McDonough School of Business at Georgetown, and although I did receive an amazing education and had a phenomenal experience, I don't recall learning about content creators and the business jobs surrounding that industry. Why is it that content creators are more effective with marketing? Because, to consumers, the content they produce feels more real, as though you are getting genuine advice rather than a company pushing you products.

"Ninety percent of consumers say that user-generated content influences their buying decisions."

—JONAS MUTHONI (MUTHONI, 2020)

Content creation has evolved into a business model. The world is moving at such a fast pace, and even the top schools may have not realized it yet, but the content-creator business model should be talked about and introduced in business school curricula. Don't get me wrong. Schools are already moving toward this direction; for example, I did have the opportunity to take a new social media marketing course while in school. The course, however, focused on social media marketing for large companies. Schools are moving in the right direction, but now I feel it is time to fully include the content-creator business model into the teachings. Sure, some people won't agree, but that's the case with any revolutionary period: eventually everyone has to adapt. Just look back at when the internet was first introduced. We have all the signs and proof that we need, which you will see in greater detail as you read this book. The earliest adopters usually get rewarded.

Throughout the years, we have traditionally experienced content creation in the form of TV shows, movies, and books. The people who created these forms of content were known as screenwriters, directors, producers, actors/actresses, and authors. Breaking into these industries wasn't easy either, and the people who managed to sneak in were part of a select few and are some of the biggest celebrities. Fast-forward to now, and we have entered a new era of content creation that breeds unorthodox creators, a term to describe the unique new-age content creators.

Unorthodox creators are taking over the entertainment industry. Anyone can create and produce media content thanks to broad availability of improved technology and the rise of social media outlets. Without the need for approval, entertainment has become democratized.

This book is for the entrepreneurs who understand that we are now in a digital world and want to understand the opportunities for them in that space. It is for the current and aspiring content creators who want to know how they can be successful in the future that is coming. It is for the business school curricula creators and students interested in the content-creator or influencer business model. It is for young adults trying to figure out how to position themselves and their lives based on what is going on around them.

What makes this the perfect time for this book? The global pandemic known as COVID-19. It sped up the timeline for the digital world, since everyone stuck in their houses needed more content to consume.

"In the first quarter of 2020, the TikTok app was downloaded 315 million times; that's the highest number of downloads in a quarter for any app ever."

—JONAS MUTHONI (MUTHONI, 2020)

This much attention could not be ignored. Also, the uncertainty of what was to come from the pandemic was high. Were we ever going to end up back outside? No one knew, so most decided to focus on the digital world, the only thing keeping us connected during this time. That focus led to many discoveries and major improvements.

Even though this book is written in a way to be entertaining and engaging, it is also meant to be a tool for people looking to not get left behind as the world changes rapidly before our eyes. This book highlights the opportunities that come from this new digital space we are entering. You will be able to learn from entertaining stories about some of your favorite unorthodox creators such as Mr. Beast, Logan Paul, Adin Ross, and many more.

Buckle up as we explore the past, present, and future of this digital world.

PART ONE

HOW WE GOT HERE

CHAPTER ONE

THE RISE OF SOCIAL

"Social media is here. It's not going away; not a passing fad. Be where your customers are: in social media."

—LORI RU

Do you remember your first social media account?

I remember mine—a Myspace account I created when I was about nine years old. As you can probably imagine, I did not have permission from my parents to be on Myspace. I knew they wouldn't want me on the platform due to my age. In fact, the guidelines specifically stated that you needed to be sixteen years old to register for the site, so of course I just didn't ask and kept it a secret.

My generation (for reference, I was born in the late nineties) happened to get to experience the birth and growth of social media. Yes, when it all started, I was a little on the younger side, but I had a lot of older cousins, a working computer, and a ton of curiosity. It was a winning combination for exploring the new world of social media. After Myspace, I moved on to so many different platforms.

At one point, it felt like every time you turned around, a new platform with different features appeared. I created so many different accounts with so many sites.

When I was thirteen, I created an account for Facebook. This was back in 2010 when, at the time, it was a relatively new site, around for about five years, and had recently expanded its reach to more than just college students. That was the first platform where I added all my classmates and family members and started sharing real moments of my life online. In fact, my first pictures that I posted to Facebook are still up there to this day. I sometimes go back to reminisce and also to marvel at the platform that Facebook has evolved into.

Social media has become such a huge part of all our lives that it's hard to picture a time before it existed, but when we really think about it, the earliest social sites came about in the nineties. Social media is truly not that old, which makes sense because the World Wide Web is not too old either, debuting just under thirty years ago in 1993. Social media sites seemingly started being created almost immediately after the internet became available for people. By the time we reached the year 2020, over 3.6 billion people were using the social media platforms all over the world (close to half of the world's population), and that number is projected to keep growing (Statista Research Department, 2022).

Let's explore how we got to that point by looking at some of the most noteworthy social media platforms or the ones that still exist today.

SIXDEGREES.COM

Six Degrees.com is considered by most as the very first social networking site, founded by Andrew Weinreich in May 1996 (Ngak, 2011). It got its name from the "six degrees of separation" theory, which posits everyone in the world is connected by no more than six degrees of separation. In 1997, the website was launched, and it included features that we will see with many other social networking sites later on, such as profiles, friends lists, and school affiliations. Even though it was the first of its kind, SixDegrees.com still amassed millions of users, actually hitting 3.5 million users at its peak (Jones, 2015). Unfortunately, one of the reasons for its downfall was the lack of people on the internet at the time, because networks were very limited. In 1999, YouthStream Media Networks purchased SixDegrees.com for $125 million, and a year later the company met its end (Bloomberg, 1999). This may have been the end for SixDegrees.com, but it would be the start of something much bigger. Many companies would later follow along in its footsteps attempting to create social networking sites.

FRIENDSTER

Next came Friendster, founded in 2001 and launched a year later in 2002. Similar to SixDegrees.com, users were allowed to sign up with their email addresses and had the ability to make friends, share videos, pictures, and messages. You could make people a part of your network, and with that, they had the ability to leave comments on your profile. Friendster caught the attention of many people and amassed over three million users within the first few months after launch, and eventually the site would even reach one hundred million users (Jones, 2015). Over the years, many new social platforms that would take over and outshine Friendster came in to play. With so much competition and so many copycats, Friendster was even at risk of other platforms poaching their users.

Jonathan Abrams, one of the founders, describes this as "a really weird time," especially because of how small they were at the time. According to him, "We viewed ourselves as the David, not the Goliath. There were the Yahoos and the AOLs, and we were this tiny, little startup. But the moment Friendster got the publicity, people started to copy it. Of course we were aware of all of them" (Fiegerman, 2014). They were also able to recognize early on that the biggest or most unique threat was Facebook, one of the social platforms that will be mentioned later in the chapter. It would seem that their concerns were warranted because by 2011 they had relaunched themselves as a gaming site, and a few years later they shut down from being beat out by the competition (McMillian, 2013). Even though social media is about people connecting, it is also about attention and staying relevant, something the next platform managed to do very well being the only one of its kind.

LINKEDIN

LinkedIn is a first-of-its-kind social networking site that focuses on professional networking. Founded in 2002 by Reid Hoffman, the site is still going strong. Like many of you, I have had personal experience using the platform.

When I got to college, finding one of my peers not using LinkedIn was almost impossible. This is where you were supposed to display your academic and professional accolades in hopes of securing internships and jobs. LinkedIn also happens to be a great place to make professional connections. This platform is different from the others because it is not purely for recreational purposes; profiles resemble résumés, which is a viable way to further your career. People saw the value in this platform from the beginning, and LinkedIn saw about 4,500 users in its first month.

By 2015, less than fifteen years later, the platform had over 575 million registered users (Jones, 2015). Currently, LinkedIn is considered the world's largest professional network containing over "756 million members in over 200 countries and territories worldwide" (LinkedIn, 2022).

MYSPACE

Soon after, Myspace launched in 2003, and of all the original social networking platforms Myspace was the most popular and likely the most influential. When it launched, it quickly grew in popularity, so much so that I even had created a profile out of curiosity as a nine year old. What started out as a file-storage platform would become a social network that allowed users to share new music. The site was clearly here to stay, and it even started getting attention from other, bigger companies. News Corporation bought Myspace for $580 million in 2005 (Jones, 2015). By 2006, Myspace would surpass Google as the most-visited website on the planet (Maryville University, 2022).

Myspace continued to be successful after the sale, generating around $800 million dollars of revenue in 2009. Shortly after, the decline began, and many people have different opinions as to why that was. One reason that we can all agree had a bit to do with it is the rise of Facebook. Then in 2011 Justin Timberlake bought Myspace in partnership with Specific Media Group. The site which once pulled in almost one billion dollars of revenue a year, was purchased for thirty-five million dollars. Now Myspace clearly isn't what it once was, but it is far from dead. It no longer has the active user base it did when it peaked in 2008, but it does pull in millions of monthly visits (Moreau, 2022).

META (FORMERLY KNOWN AS FACEBOOK)

Now we get to perhaps the most iconic social networking platform, Facebook. In 2004, Mark Zuckerberg, the Harvard dropout, as well as Eduardo Saverin, Andrew McCollum, Dustin Moskovitz, and Chris Hughes founded the site that gave many other founders nightmares. The platform started as a site that was exclusive to Harvard students but then quickly spread after that. By 2006, the site was available to anyone thirteen years or older. Facebook became a way to connect with family, friends, or old school mates. Through Facebook, grandparents got the opportunity to see pictures of the kids. Facebook was genius concept that developed into so much more. The site got so popular that, by 2008, it surpassed Myspace and became the most-visited website in the world (Jones, 2015).

Since then, Google has reclaimed the throne as the most-visited website, but Facebook remains number three (YouTube holds the number two spot) just to give you a glimpse of its longevity. In 2012, Facebook had an initial public offering of $104 billion, one of the highest initial public offering valuations ever. By 2015, the site was generating over forty billion dollars a year in revenue (Jones, 2015). Facebook currently also has 2.85 billion monthly active users (Statista Research Department, 2022).

YOUTUBE

One of the most iconic social platforms today, YouTube registered as a site in 2005 by Steve Chen, Chad Hurley, and Jawed Karim. All three of the founders were employed by PayPal before deciding to start the new platform. The idea originally formed with the intention of being a video dating platform. In order to get people on the site at first, they even offered twenty dollars to women to post videos of themselves to attract users, but no one came forward. Soon, they realized it wasn't working, and they needed a new approach. That's when one of the founders, Chen, proposed a new idea, "Okay, forget the dating aspect. Let's just open it up to any video."

The first video uploaded to the platform was Karim's video, "Me at the zoo," an eighteen-second video of elephants. Chen's idea would lead to something legendary. After the first video, the site saw extraordinary growth, so much so that Google would purchase it for $1.65 billion in 2006 (Dredge, 2016). The growth has been exponential ever since; YouTube now has over two billion logged-in users every month and is the second-most-visited site on the web behind Google (YouTube, 2022).

REDDIT

Much different than the platforms listed before it, Reddit was launched in 2005 by Steve Huffman and Alexis Ohanian. The site launched out of the Y Combinator program where the twenty-two-year-old founders were able to secure one hundred thousand dollars in funding. The site initially started as a news-sharing platform. A year later, they were pulling in about five hundred thousand unique visitors a day, which led to Condé Nast Publications purchasing Reddit for twenty million dollars (Moradian, 2020). Reddit later transformed from just a news-sharing platform into a combination of news aggregation and social commentary. The main feature that people loved was the ability to "upvote" and "downvote" (Maryville University, 2022). The founders left after a few years to work on new ventures, but the platform continued to grow. Reddit currently has fifty-two million daily active users worldwide (Lin, 2021).

TWITTER

The next platform, Twitter, was created in 2006 by Jack Dorsey, Noah Glass, Biz Stone, and Evan Williams. The platform had an interesting concept, allowing people to make posts but limiting them to only 140 characters, basically creating a sort of microblogging platform. In 2013, Twitter had an initial public offering valued at $14.2 billion. The platform is also iconic in a way because some celebrities tend to be quite active on it, engaging users by sharing parts of their lives with followers and the world, like the random rants by Kanye West or the occasional flex by a Kardashian. In 2017, the platform doubled the 140-character policy in what I can only assume was a move to give users more space to create (Jones, 2015). As of 2021, Twitter has 192 million daily active users on its platform (Lin, 2021). It has even evolved into a place where people look for their daily news.

INSTAGRAM

Kevin Systrom launched Instagram in 2010. The idea for this platform was based around photos captured on your mobile device. With cameras getting better and better, the platform had so much potential and quickly grew in popularity. Instagram gained twenty-five thousand users in one day, and by the end of the first week, had one hundred thousand downloads. Within the first three months, the platform had reached one million users. By 2012, it had grown to twenty-seven million users, becoming so popular that Facebook would make an offer of one billion dollars to acquire the platform, an offer that was accepted (Blystone, 2020). As of 2021, Instagram has over 1.07 billion users worldwide, a number that will continue to grow (Mohsin, 2021).

SNAPCHAT

The next platform, Snapchat, had a very unique concept: posts that only live for a day, a concept which people loved. You can post pictures or videos to your story, but they disappear in twenty-four hours. The platform was founded in 2011 by three Stanford students—Evan Spiegel, Reggie Brown, and Bobby Murphy—and was originally named "Picaboo." A few months later, Reggie was forced out of the company, and Evan and Bobby relaunched the company with the name Snapchat. The company grew, and new features were added that would make users' experiences better, such as geofilters and Snapcash. By 2015, Snapchat had seventy-five million monthly users worldwide, and advertisements accounting for 99 percent of its. In 2017, it had an initial public offering of twenty-five billion dollars (O'Connell, 2020). As of 2021, Snapchat had reached 500 million active monthly users and 280 million daily users (Rodriguez, 2021).

TWITCH

Twitch was founded in 2011 by Justin Kan. Twitch, however, has a history that dates back to 2005 under a different name, Justin.tv, a place that allowed people to livestream, giving off somewhat of a reality-television vibe because of the content. Back in that time, streaming video games was difficult, so the focus then became getting people to stream their lives, creating the term "livecasting." When the platform Twitch actually launched in 2011, it was fully focused on e-sports and gaming. At launch, it was pulling in 3.2 million unique visitors per month, and a year later in 2012, it had twenty million unique visitors per month

(Cook, 2014). By 2014, many companies became interested in purchasing the platform. In 2015, Amazon purchased Twitch for $970 million (Epstein, 2021). In January 2021, the platform had two billion hours of watch time; yes, people had actually consumed two billion hours of content from this place (Iqbal, 2022).

TIKTOK

Perhaps the most iconic platform is TikTok, founded in 2016 by a Chinese tech company called ByteDance Ltd. Ever since, TikTok has been growing at such a fast pace. It launched at the end of 2016, and by midway through 2018, the platform had reached half a billion users. The platform is about posting short-form content: fifteen-second, sixty-second, and, most recently, three-minute-long videos. By 2021, the app had been downloaded over two billion times and had 689 million active users worldwide. The platform has been the top-downloaded app on the Apple App Store for over five consecutive quarters (Mohsin, 2021). It's growing at a pace much faster than any of the previous platforms.

One thing common among all the social media platforms mentioned in this chapter is that they all attracted a lot of people; each one of them had millions or even billions of users. This is primarily due to the fact that people want to interact and connect with each other. As more and more people continue to get on the internet, the amount of users on social platforms will continue to rise, which will lead to many more opportunities. We've already started to see the shift and effect that the rise of social media has had on the business world; we see influencers getting brand deals, as well as companies running ads on these social platforms. This will continue happening as the general public continues migrating toward social media platforms.

Based on the information in this chapter, social platforms clearly aren't going anywhere, so it's best we explore what that means for the future. What are the business opportunities? What are the career opportunities? What does this mean for content creators? How do we survive in this time? Not just survive, but how do we thrive in this time? What does it mean for our youth?

The short answer is that we need to adapt, and as a society, we will as we always have throughout the years. But how do you make these changes work for you? You make them work for you by being the unorthodox creator. Later in the book we will explore just how this can be done.

CHAPTER TWO

BARRIERS REMOVED

"Falling entry barriers and lower access costs have significantly democratized participation, whether in production or consumption."

—MOHAMED EL-ERIAN

Once upon a time, in order to get a good picture of yourself, you had to go to a professional photographer or at the very least someone with a professional-grade camera. Once upon a time, to create really awesome videos—whether movies, TV shows, short films, or even just marketing promo videos—you needed to hire a production team and camera crew with professional-grade equipment. Once upon a time, if you wanted to do some writing for a book or even a blog, you needed to go to the library to use a computer or own one yourself. One of the greatest parts about living in the world today and not once upon a time is that the average person can now do all these tasks from pretty much anywhere, and it's all because of one great invention.

What do we all own now that most people didn't fifteen years ago?

If you guessed a smartphone, you're right, but at this point it's much more than just a smartphone: it is the key to the internet and your magical gateway to being a creator all in one. I remember taking two-megapixel photos with cell phone cameras back in the day to now creating full 4K YouTube videos with just my cell phone. Even with my first book, I definitely wrote significant parts from my cell phone. What kind of cell phone do I have? you might ask. The iconic Apple iPhone.

At Macworld 2007, Steve Jobs revealed the first iPhone, a new-concept touchscreen phone that would eventually change the way of the world. It was a silver-and-black metal-back phone with a glass screen. It's key features were a 3.5-inch diagonal screen, 320 x 480 pixels at 163 ppi, and even a two-megapixel camera—a little, pocket-friendly device with endless possibilities (Montgomery and Mingis, 2021). The phone launched in June 2007, and many companies would create their own versions of what would be known as "iPhone killers," but it wasn't until 2012 that it became the norm to walk around with one of these devices. With the internet already taking off, these smartphones were like an add-on, a way to amplify what we could do. A tool that we could at one time only use from the comfort of our homes was now available for us to use at any time, and all it would take is reaching into our pockets. With the growth in the number of people using these devices, programmers got to work and started creating mobile applications. Now this may seem obvious, but mobile devices have completely changed the way we experience life. There is an app for almost anything, and we have so much information and utility available to us. Whether it's ordering a meal to your door, checking your bank account, or filming a high-quality movie, we have the power to do all of this at our fingertips.

As the years have gone on, these devices have only improved all while getting more and more affordable, becoming little super machines that most of us walk around with in our pockets. Most of these phones can record not just HD but also 4K, and a lot of them also have the ability to record in slow motion. These phones are essentially the professional-grade camera that you need to jumpstart you career on social media when you might have no other resources or extra money.

What does that mean? Is it possible for anyone to be on social media? Even someone who may be homeless?

Well, the answer is yes, and one man proved it.

The man goes by the name Oneya Johnson, though you may be more familiar with his TikTok name @angry-reactions (Chen, 2022). Based out of Lafayette, Indiana, he was twenty-two years old when he created his page. When he created the page, it was as a joke, and he would present an angry character spreading positivity through reactions to other peoples' content. His first post on his page was a video of him reacting to another user making a cake and "angrily" saying how much he liked the cake and other positive things (D'Amelio, 2022). He came up with the concept for his videos based on societal views of himself.

He said, "I basically took how the world sees me, and how I really am, and just matched it together. If I passed you on the street, and I don't say a word, I look like the angriest person in the world. But when you actually get to know me, I'm actually a really positive person." After posting that first video, he blew up overnight, gaining over one million followers within twenty-four hours. He didn't expect that would have come out of it, but who could blame him?

> *"I honestly couldn't believe how fast that account blew up. I still can't believe it."*
>
> —ONEYA JOHNSON (CHEN, 2022)

That's not the craziest part though; what is even wilder about the entire situation is that he was homeless at the time, living and filming these videos out of his car. He had previously shared an apartment with his ex-girlfriend, Jillian, in Michigan before he was evicted by their landlord after countless noise complaints by the neighbors due to arguments he had with his ex-girlfriend. No charges were ever pressed against him, but it did lead to the eviction. Jillian confirmed this, "Oneya is big, and his voice is bigger than mine, so neighbors could only hear him." She also confirmed they are on good terms today (Chen, 2022).

His TikTok page now has more than 22 million followers and has managed to rack up over 404 million likes. He managed to secure brand deals and now also sells merchandise to promote his page, and he started all of this while homeless, something that would not have been possible before smartphones were available to everyone.

Oneya was not the only homeless guy who took advantage of having a smartphone and the magical app known as TikTok. Another TikToker by the name Zeemer, better known by his username @randomhomelessguy2, lives with his mom in their car and posts videos about how he prepares his food as a homeless man (Sungailaite, 2021). After only doing this for about year at the time of my writing, he has managed to gain 2.9 million followers and 46.4 million likes, something else previously not able to happen before smartphones. He always tells his viewers that he does not want them to send him any money. Based on his new videos, it appears that he is still homeless or continues to make this type of content because it's what brought him the notoriety. Either way, he has definitely gotten to a point where he could be getting funded for his views. Both creators' new situations would not be possible without the progression of technology.

Homeless people turned social media stars are not the only people the improvements in technology helped. Improved technology has also affected education in a major way. When it comes to researching anything, a majority of us now have laptops that can connect to the internet and find what we're looking for in a matter of seconds. This makes completing projects and assignments so much easier. Think about it: Depending on when you grew up, to do this, you had to venture to the library. Just ask your parents. Then you had to think about if certain neighborhoods had libraries or not. Students who were in neighborhoods that had libraries had advantages over students who were in neighborhoods that didn't have libraries—just an example of how the rise in and progression of technology started to bridge gaps in society.

Improved technology, especially in the consumer space, helps to bridge the socioeconomic gap especially. It doesn't just stop with being able to research; we've also gotten to a point with the internet where we have so much content online through social platforms (which you saw in the previous chapter), including a lot of educational information. Many people use improved technology to post tutorial videos of a wide range of topics to sites like YouTube. Maybe a child can't afford to pay a tutor for help with algebra, but they most likely have access to the internet where they can watch video tutorials on how to solve some of the equations.

This is not where it stops though. The benefit of the improvements carries over into all facets of our lives, like our careers. Improvements in technology have given people the opportunity to work from home for a major company or for themselves. Someone who would once upon a time have to venture to the office every day now has the opportunity to stay at home and still make an income. Now this sounds similar to what content creators do, but the difference is that you didn't need to create your own work; you're not taking on the risks of working for yourself, as you still are working for big companies and have the security of a salary. This idea of working from home has been possible for a while now, but the pandemic truly made people realize that it could work on a huge scale. This gave people the opportunity to raise their children and take care of their families. This gave the world a way to keep moving forward during a global pandemic, as no one knew what was going to happen or when this tragic event was going to end, but some people were given the ability to continue working to support their families from home through this time. Through this time, many companies have also seen that in some cases the improved technologies have gotten to a point where employees do not lose productivity by staying home, leading to the company saving money as well.

These improvements in technology have also led to people being more creative and people creating their own opportunities for themselves. Why just turn in a regular application to a competitive program when you could give yourself a better chance by creating a potentially viral video? The possibilities are endless. I've witnessed little girls selling Girl Scout cookies online. There is no limit to what people can do with the rapid growth of technology, and it will continue to lead to improvements in people's lives in a major way, from financially to educationally to even recreationally.

The rise in technology paired with the rise in social media, as well as access to internet becoming available to more people, truly removed the barriers for content creation. As technology improved, prices lowered, and it wasn't just for the upper class but for the mass public. This changed a lot by giving people the opportunity to break into the media space. If someone had raw talent and could entertain people, success would follow whether they had a setup worth thousands of dollars or if they were homeless with just a phone, like the amazing creators we talked about earlier. For the creators who didn't have much, their content is most times the key to changing their lives and standards of living. As long as you have access to a device that can connect to the internet, you have the opportunity to change your life.

CHAPTER THREE

WHAT IS INFLUENCE?

"Think about what people are doing on Facebook today. They're keeping up with their friends and family, but they're also building an image and identity for themselves, which in a sense is their brand. They're connecting with the audience that they want to connect to. It's almost a disadvantage if you're not on it now."

—MARK ZUCKERBERG, COFOUNDER
AND CEO OF FACEBOOK

We have arrived at a point in time with the rise in social media where people are developing images of themselves online. Based on the image that people create for themselves, some of them have gained influence and developed followings now making them influencers. The social media influencer is anyone with a following online; it doesn't even need to be a super large following, as just having a few thousand would make you a micro influencer. The image influencers create also doesn't necessarily need to be positive or negative; it just needs to resonate with people and make an impact, and they just need some people to care. There are many positives and negatives to this. For example, motivational influencers have gained a following by uplifting their communities, but on the other hand, some negative influencers have built a community around hate.

The important trait that influencers contain is the ability to make others feel. It doesn't matter what their followers feel but more so how much these influencers can channel these emotions. There are so many benefits to having that ability, but the biggest benefit is the potential to create a following or grow your social currency. This leads to having the power to influence people's purchasing decisions. That power of influence online today is so valuable that companies are willing to pay anyone who has it to promote their products. The main factors that lead to this online influence are content, expertise, attractiveness, social identity, and trust (Gashi, 2017).

The amount that an influencer is paid is not public knowledge, and the rates vary based on the influencer's following and negotiation skills. Someone did give some insight into what some of these rates look like. David Dobrik, a famous social media influencer who is most known for his YouTube vlogging videos, said on his podcast Views that someone with one million followers and about a 20 percent engagement rate should get about fifteen thousand dollars for a YouTube video, about two thousand dollars for an Instagram story post, and about five thousand dollars for an Instagram feed post (Dobrik and Nash, 2020). It is important to remember, however, that these numbers are not set in stone.

The thing with social media influencers today in the digital world is that they could focus on whatever they want. Whatever their passion may be, they can create content surrounding it. So many consumers are now on these platforms that it is very likely there is a community of people who would resonate with content about anything. That has led to people becoming music influencers, art influencers, gaming influencers, technology influencers, makeup influencers; pretty much any topic you can think of already has influencers in the space. From the broadest topics down to the most niche ones like a realistic tattoo artist influencer.

One man foresaw this idea before a lot of other people, and because of this genius foresight, he is regarded as one of the gurus of social media. This man's name is Gary Vaynerchuk.

Gary Vaynerchuk is an immigrant who was born in Babruysk, Belarus, in the former Soviet Union. He moved to the United States in 1978 when he was three years old. He grew up in the tristate area and lived with eight family members in a studio apartment in Queens, New York, before they later moved to Edison, New Jersey. He was always an entrepreneur at heart, starting a successful lemonade business at the age of seven years old, and then selling thousands of dollars' worth of baseball cards and toys by the time he was in high school.

Although he had an entrepreneurial mind at such a young age, this knowledge didn't translate to school. Gary actually talks about this topic (how bad of a student he was, consistently getting bad grades) a lot, for motivational purposes, of course. It is interesting though because he wasn't a bad learner; he just didn't perform particularly well in the classroom. But he did learn quite a bit from his entrepreneurial ventures, such as the supply-and demand-model that will benefit him forever.

Gary came from humble beginnings: His parents didn't come to America with a lot of money, but they worked hard to change their situation. His father, Sasha Vayner-chuk, worked long, hard hours as a stock boy in a liquor store. After a while working hard in that role, he was promoted to assistant manager of the store and later was promoted to the actual manager. During these years of working hard, he was saving money, enough to eventually start his own business. Sasha took his savings and purchased a liquor store in Springfield, New Jersey. Little did he know at the time but that liquor store would lead to the rise of the internet personality, Gary Vee (Vayner-chuk, 2022).

When Gary was in high school, he began working for the family business in the liquor store, bagging ice. As he got older he transitioned into more of a salesman-type position. One day while working in the store, a few customers came in looking for a particular wine, but it was unavailable at the time. Keep in mind this was during the nineties, so a backorder system didn't exist besides Gary manually taking down the names of people interested in the unavailable wine. This frustrated him. Then he experienced something very interesting: a wine collector came into the store and backordered six cases of a particular wine. This brought him back to his baseball card collector days when he would sell to collectors, and he realized that he could sell to wine collectors the same way. He then made it his mission to study up and become an expert on wine (Reddy, 2021).

Something interesting happened while he was in college. He was introduced to the internet, and when he was first exposed to it, he stood in front of the computer and watched for hours. After doing some more exploring, he learned that people could sell things using the power of the internet. He learned about e-commerce platforms like eBay and Amazon. In 1996, he transformed the family business by creating WineLibrary.com, one of the first e-commerce wine businesses in the United States. Yes, this does mean that eventually they changed the name of the business to "Wine Library" (Reddy, 2021).

In 1997, he created an email newsletter. During this time, many people didn't even have email addresses; however, a year later his newsletter was being sent to two hundred thousand emails. In 1998, he graduated from Mount Ida College and immediately went full time into the family business where he would take over daily operations. He worked really hard to transform the business that was mainly an offline liquor store to an online wine business. In 1998, the business was pulling in three million dollars in revenue a year with 10 percent of that being profit, and by 2003, he grew that revenue to sixty million dollars a year. He focused a lot more on online and digital marketing versus the traditional forms of marketing (Vaynerchuk, 2022).

In 2006, Gary started a YouTube channel, which started the path to him becoming a huge influencer. The name of the YouTube channel was WineLibraryTV, and he produced a new video every day for five years, something that grew his following and started to build his popularity (WineLibraryTV, 2022). Out of this buzz came an invite to Conan O'Brien's show in 2007, which produced a clip of him that went viral on YouTube (Rebecca Brooks, 2009). This clip led to him becoming one of the most followed people on Twitter at the time. His rise in fame just led to more and more television appearances such as on *The Today Show* and *The Ellen DeGeneres Show*, all of which grew his social following even more. This then led to bigger companies reaching out for Gary's expertise. He created multiple businesses out of this following and even wrote and sold multiples books with the power of his influence (Reddy, 2021).

Gary was one of the first people to explore the Web2 space and become an expert in it. The Web2 space is essentially the second phase of the internet that focuses on user-generated content.

"The second stage of development of the World Wide Web, characterized especially by the change from static web pages to dynamic or user-generated content and the growth of social media."

—LEXICO.COM DICTIONARY (LEXICO.COM, 2022)

His early expertise led to the power and influence he holds today. He is not the only person to do this though, as many other early adopters have managed to create massive followings. Social media influencers are known for a wide range of topics online, and companies use that as an opportunity to get their products or services sold. The opportunity is a lot bigger than most would originally expect, but if you have been paying attention to the trends, you see it too.

In today's time, social currency is proving to be more valuable than ever, and as we are seeing brands and companies reevaluating how to dish out their marketing budgets, we are also seeing influencers create their own products and brands. Some companies have even partnered with influencers to build brands which you will see examples of later in the book.

Understanding the concept of the influencer and influence is important in understanding the direction in which this new digital world is going. The speed at which this digital world is growing is allowing the importance of the influencer to grow.

"It took thirty-eight years before fifty million people gained access to radios. It took television thirteen years to earn an audience that size. It took Instagram a year and a half."

—GARY VAYNERCHUK (VAYNERCHUK, 2013)

The idea that the power these influencers hold is valuable remains true, but I believe it is more valuable than we currently appraise it. If the digital world continues the pace it is on, influencers will continue to gain more power and influence. Promoting products remains a great job for these influencers, but later on in this book we will explore just how much further we can take that idea. It isn't just about the promotion of products, however; influencers influence people's life choices, financial decisions, and political decisions. As this continues, we will start to see influencers take on more important roles in society in addition to more people in society becoming influencers or internet personalities—because at this point, in order to stay relevant and have validity, there's no way around it.

PART TWO

PRINCIPLES OF OLD-MEDIA VERSUS NEW-MEDIA MINDSETS

CHAPTER FOUR

ONE YES VERSUS ONE MILLION

"Social media is the democratization of information, transforming people from content readers into publishers. It is the shift from a broadcast mechanism, one-to-many, to a many-to-many model, rooted in conversations between authors, people, and peers."

—BRIAN SOLIS

Let's say you had an idea for a really cool project, maybe a movie or a TV show of some sort. A few years ago, the only way to get a project like this out to the public was to go to a broadcasting channel to make it happen for you. That means that you would have to convince a handful of other people to understand your vision in hopes that they would make it a reality. What if they didn't like it? What if they didn't like you? What if they didn't see the vision as clearly as you did? What if they wanted creative control?

What if instead of going through all of that, you could take your idea straight to the masses? Social media gives you the opportunity to go straight to the people. You no longer need the middleman. You can take your ideas, your art, your work straight to the consumers, and they can decide whether they like it or not. In this new digital world, that is exactly what some people decided to do, and it worked out for them really well.

You no longer have to deal with the machine if you don't want to, the machine being the big companies whose approval you previously needed in order to show your art, publish your work, or display your talents. The machine refers to the production companies, the publishing companies, the television networks, the movie studios, etc.

At one point in time, before the digital world and technology progressed so far, people in creative spaces had a desire to make it to these machines and be accepted by them. This new time has changed things drastically. For example, an aspiring actor previously would have to fill their schedules with auditions and hope to get a gig; today an aspiring actor has the ability to pick up a camera and show their skills to the world through platforms like YouTube. An aspiring article writer previously would have to pitch their ideas to newspapers or magazines in order to get their work out to the masses; now an aspiring article writer can post their work online to blog sites.

This digital world allows you the opportunity to bypass the prejudice that some people may have against you. Actors and actresses have often complained about experiencing some of these prejudices; even some of the biggest stars have experienced it.

> *"You would hear a lot, Black people don't translate internationally."*

—WILL SMITH (MALIVINDI, 2020)

Will wasn't the only star to experience this prejudice though. Many actors, performers, and entertainers have experienced similar things. Another great example of this comes from Asian Hollywood actress Lucy Liu. She experienced quite a bit of prejudice when it came to even getting auditions for roles.

> *"I had some idea when I got to LA, because a friend of mine would have ten auditions in a day or a week, and I would have maybe two or three in a month, so I knew it was going to be much more limited for me."*

—LUCY LIU (MALIVINDI, 2020)

Imagine that. She couldn't even get auditions for roles, and you can't play a role if you can't even get in the door. It was so bad that a lot of agents didn't even really want to work with her, not because they didn't like her but because they weren't up for the challenge.

"Everyone was willing to have me on their roster but not commit to me because they didn't know, realistically, how many auditions I could get."

—LUCY LIU (MALIVINDI, 2020)

With the digital world, Lucy and other actresses wouldn't have to deal with this. She could act in her own videos and post them online without the need for auditions or even the agents.

The beauty of the digital world is that it doesn't mean that the machines are now obsolete; it just means that the creators now have more power to control their own futures. Some creators have even used their success in the digital world on these social platforms as leverage with the machines. It's normal today to see a popular online content creator end up with a role in a movie or guest starring on a television show. Even aspiring author Laura McVeigh was able to secure a publishing deal through the power of Twitter (White, 2016).

Now we'll explore just how easy this new digital world made it for creators to take matters into their own hands and just create without having received the traditional cosign that was previously needed. With the technological advances in consumer products like cell phones and improvements that we've made to social media platforms, it's now very easy to do this, as we learned in chapters one and two.

The first person we'll look at is the legendary writer, actress, and producer Issa Rae. Coincidentally she was born in Los Angeles, the city of show business where multiple movie studios reside. Her family moved to Potomac, Maryland briefly before moving back to Southern California when she was in sixth grade. Her father was a doctor, and her mother was a schoolteacher, which is probably part of the reason she ended up at King/Drew Magnet High School of Medicine and Science. While attending that high school, she became interested in theater and even managed to secure the lead in the school play for all four years. She would study movies and music videos and learned pretty soon after that because of her race, she would always have to be a side character.

> *"You learn that a very specific type is appreciated. For me, it was like, 'If I want to pursue acting, I know that I am going to always have to be the best friend.'"*

—ISSA RAE (NWANDU, 2018)

She then went on to Stanford University where she majored in African American studies with a minor in political science and graduated in 2007. Issa started creating a web series that she would post to YouTube, casting others as the lead until she just decided to cast herself. The name of the now iconic YouTube series is *The Mis-Adventures of Awkward Black Girl* (Rae, 2011). She dealt with doubt and embarrassment, at the time still not confident as to where her career in entertainment was heading (Nwandu, 2018).

> *"The embarrassment came from making a YouTube series while all of my friends were being doctors, lawyers, diplomats, all of those different things. Those postcollege questions—did I have to go to college to do this? Did I have to pay hundreds of thousands of dollars to make YouTube videos?—that was embarrassing for me."—Issa Rae (Nwandu, 2018)*

Eventually the doubt would go away when she started making money, and her buzz also grew which led to her getting multiple television and movie opportunities. She is most likely best known for the show *Insecure*, a show she wrote and starred in (Rae and Wilmore, 2016). This truly shows that because she posted her work to YouTube and gave people the choice to choose her, she also unlocked the support of major production studios.

She's since paved the way for many creators, but one group in particular that she's helped is the YouTube group Rdcworld1, currently some of my favorite content creators online today (RDCworld1, 2022). They are led by Mark Phillips, who many would agree is sort of a genius in the content-creation space with his small skits. Much like Issa, Mark also has created his own unique series on YouTube, like *Anime House* and *Videogame House*, both shows that feature some of your favorite characters from anime and videogames living in a house together—brilliant ideas that have garnered a ton of attention from millions of people (RDCworld1, 2016 and 2019). That hard work has paid off, and now Rdcworld is getting two different shows on HBO; and you won't be surprised to know that Issa had something to do with that.

Another creator who took matters into their own hands is YouTuber turned music artist Darryl Dwayne Granberry, Jr. from Pontiac, Michigan, but he is best known by his stage name DDG (DDG, 2022). He was a pretty smart student and graduated valedictorian of his high school, International Tech Academy, which then led to him attending Central Michigan University. While in college, he started a YouTube channel, and it wasn't long until he would drop out and move to Los Angeles to pursue his content-creation career with no additional workload. Like most people when they decide to drop out to follow entrepreneurship, DDG did so because the money started coming in. Throughout his YouTube career, it would appear that he had a passion for creating music, so he did that while building a following from his vlog-style videos (Aade, 2021).

A few years ago, I would've said that it was pretty difficult to break into the music industry because of the need for a label; but in this new digital world, you could take the product straight to market, and that's exactly what DDG did. Leveraging the following and community that he developed, he was able to drop music that would automatically get streams due to loyal fan base. Then there came a point in time on YouTube when diss tracks became a popular song-creation trend, which led to more people listening to DDGs music when he participated. As he grew in popularity he, was able to then work with more prominent people in the music industry which led to more streams of his music. It got to a point where his music on YouTube was pulling in millions of views (Aade, 2021).

Eventually his success in the music scene became be too much for record labels to continue ignoring, and in 2018 he inked a deal with Epic Records. With the success he saw from music, he decided to put more time into music and less into making vlogs. This led to a platinum record as well as a spot on the 2021 XXL Freshman Class list, one of the most sought-after designations as a new rapper (Cline, 2021). DDG was able to become a mainstream rapper because of this new digital world.

The democratization of content is truly changing the media world, affecting both traditional media and new media. This is one of the reasons we now have so much content to choose from. It also means that people no longer need to wait or have approval, and they don't need a group of executives to see their vision first. Democratization gives the content creators more power in terms of creativity and control over what they release to the world, but more importantly the consumers of the content hold the power. The traditional media is paying attention, though, as you can now sometimes see your favorite content creators casted in traditional movies or television shows. This concept will continue evolving in the future, and I envision it with creators gaining even more ownership, especially as we start to venture in the Web3 space. We will learn more about that later on in the book.

CHAPTER FIVE

ALGORITHMS AND COMPLEX EQUATIONS

"The brain is the most complex, challenging scientific puzzle we have ever tried to decode."

—PAUL ALLEN

Yeah, the brain is complex, and so are the algorithms for social platforms.

With the rise of social media platforms and the increase in the success that users are finding, more and more of us are venturing on to them. More people on social media means more competition, but to the platforms it means a ton more content to sort through and present to their users, which brings us to one of the most important yet complex parts of social media: the algorithms.

You might be asking, what are social media algorithms?

"Social media algorithms are a way of sorting posts in a users' feed based on relevancy instead of publish time.

Social networks prioritize which content a user sees in their feed first by the likelihood that they'll actually want to see it. Before the switch to algorithms, most social media feeds displayed posts in reverse chronological order. In short, the newest posts from accounts a user followed showed up first." (Barnhart, 2021)

Now you're probably wondering why these platforms have moved away from chronological order, and after some research and deep thought, I have come to a few conclusions. These platforms moved on to attempting to create the right algorithm that would make users happy; after all, the more users you have and can retain, the better for your platform.

Content creators have been trying to figure out how these algorithms work since the time that they first arrived, and many have tried, and many have failed. Figuring out how the algorithm works would be like figuring out the pin to a safe with unlimited money. This is because on social media platforms you typically get paid more when you get more views.

One content creator from YouTube has seemed to have figured out the magical YouTube algorithm, and his name is Jimmy Donaldson, better known as MrBeast. He was born on May 7, 1998, and grew up in Greenville, North Carolina (Leskin, Russell and Asarch, 2021).

Jimmy uploaded his first video to YouTube when he was twelve years old, and his original YouTube channel name was MrBeast6000 (MrBeast6000, 2022). From that young age he became obsessed with trying to crack the code of the YouTube algorithm. Even though he probably did not know much about algorithms at the time, he was interested in how to get his content to attract the biggest audience it could, so he decided to have two channels. With his first YouTube channels he focused on posting gaming content—*Call of Duty* videos on one and *Minecraft* videos on the other (Shaw and Bergen, 2020).

As time went on, he became obsessed with the success that other creators had on the platform and the economics of the site, which led to him creating a series of videos estimating the earnings of some of the top creators on the platform, a video series that he has since deleted. He started with PewDiePie, who he happens to be a huge fan of. PewDiePie also happens to be the most-followed person on the platform today (MrBeast, 2015). He then started a few other different types of video series like offering tips and tricks to aspiring creators and commentating on YouTuber drama. He was clearly trying differentiate things to gain more viewership, and with his next series he actually did. He created a YouTube video series sharing the "worst intros," where he made some friendly jokes about other YouTuber's intros, typically the first ten to thirty seconds of their videos. He did this in 2015 and 2016, and by midway through 2016 he reached thirty thousand subscribers.

The year 2016 was massive for Jimmy. Aside from the growing following, he also graduated from high school, Greenville Christian Academy, and at the request of his mother, also attended college at East Carolina University.

That would be short lived however as he would dropout after only two weeks. A tough conversation that he had with his mother led to his statement below:

> *"I'd rather be poor than do anything besides YouTube."*

—JIMMY DONALDSON (LESKIN, RUSSELL AND ASARCH, 2021)

He continued to study the platform and work at it. Then in January 2017, he created his first viral video where he counted up to one hundred thousand, a video that took him forty-four hours to complete (MrBeast, 2017). This experience gave him a lot of insight and started the route down the path of success for Jimmy.

> *Over time, he deduced more of YouTube's mysteries. Make a clip too long, and no one watches or people want to watch another. Make one too short, and people won't linger. Use a bad thumbnail photo or title, and no one will click. Donaldson typically makes videos that are between ten minutes and twenty minutes long. He picks a concept that is easy to communicate in the title—e.g., "I Adopted EVERY Dog in a Dog Shelter"—and then uses the first thirty seconds to establish the stakes (Shaw and Bergen, 2020).*

I am writing this in the beginning of 2022, and Jimmy currently sits on six different YouTube channels with his combined subscriber count adding up to over 160.07

million, with his main channel accounting for 91.7 million of those subscribers (MrBeast, 2022). He remains one of the fastest-growing YouTubers on the platform. He may not hold the title for having the most subscribers yet, but it seems that he is approaching it at a very fast pace. At the time of writing, only one single person on YouTube is ahead of him—PewDiePie, who sits at 111 million subscribers. In terms of YouTube channels in general, Jimmy sits in the fourth spot behind PewDiePie, Sony Entertainment Television India (126 million), Cocomelon - Nursery Rhymes (128 million), and T-Series (206 million) (Feeney, 2022). He also manages to pull in an absurd amount of views per video which a lot of other top creators can't do. On his main channel, he is just shy of 15 billion views on his 719 videos.

The reason Jimmy is so successful in comparison to a lot of creators is because of his research and approach to the platform; his years on the platform have taught him what garners interest. He has mastered thumbnails and titles, and he knows what a user would want to click on. He also managed to master the perfect format for YouTube videos: preview the climax at the beginning of the video, let the users know what is coming, then show the buildup. He has also gotten to a point where he spends a pretty penny on his videos with the objective of doing the biggest and best.

With this he paved the way for new creators like the YouTuber Airrack (Airrack, 2022). What Airrack did is supposed to be impossible, but honestly with the blueprint already available to everyone, all it really takes is execution. Airrack followed the MrBeast blueprint and managed to gain one million subscribers on YouTube in a year, which is pretty unheard of. He also did something very unique. He found unorthodox ways of collaborating with already big name creators; whether it was by sneaking into events or buying a super expensive couch, he was very innovative with his approach.

As we continue to explore this new digital world, we clearly still have much to learn. However, as we learned in this chapter, geniuses may have created these platforms, but it doesn't take the same kind of genius to figure it out and crack their codes. All it takes is a person willing enough to pay attention. You don't need to be as obsessed as MrBeast was, but it doesn't hurt. Many people have adopted MrBeast's style as he is pretty open about his YouTube strategy. Still he has yet to be outperformed, primarily because many are not willing to go to the lengths that he is, whether financial or otherwise.

Even though knowing how the algorithm works is important, making great videos is going to remain the reason why people decide to subscribe or follow you. The knowledge behind how these algorithms work will prove to be very valuable not just to content creators but to anyone who wants to understand or explore these platforms. As time goes on, these platforms will continue to revise and update their algorithms in an attempt to stay on top. Being early to figuring the revisions out and adjust will be one way to grow on these different platforms. A creator can be early by staying on top of trends and paying attention to what's going on. Also, whenever a new platform shows it's head, it's a creators job to figure it out and utilize it to their benefit. The only way for creators to be early is to keep evolving with the times. New platforms are always going to appear, trying to be better than their predecessors, and algorithms will still be relevant in an attempt to always give users the content they want to see.

CHAPTER SIX

BLACKBALLED VERSUS REDEMPTION

"A good act does not wash out the bad, nor a bad act the good. Each should have its own reward."

—GEORGE R.R. MARTIN

We have seen and heard many stories by now about people getting cancelled, but we have also heard about many redemption stories. One thing you might not have thought about is the relationship to social media that these two have. To be quite frank, before social media, it was really hard for people to be redeemed after something big. I believe this is the case because before social media, people only saw entertainers, superstars, and public figures through their work. Social media has a way of humanizing someone. You realize, just like you, these people have likes, they have dislikes, they have opinions, and they have emotions. If you make your money from creating content online or have a strong community of followers, you don't really need to rely on other means for work or pay. Actually, in terms of social media versus traditional media or any other job where the person is an employee, when you do something pretty bad, you will most likely lose your job and can even be blackballed and not be able to find a job in the same industry.

A great example of this was Kevin Spacey who starred in the hit show *House of Cards* (Willimon, 2013). Spacey was dropped from the show after sexual assault allegations (Romano, 2018). I won't comment much about what happened legally in the case, except to say that I do condemn any kind of sexual assault. I am bringing this up to focus on the comparison of what happened to Spacey in a work sense and what happened to the Vlog Squad, a popular group of content creators that had one of its members accused of sexual assault.

One of the Vlog Squad members—or, at this point, I should say ex-Vlog Squad members, as all the other members cut ties with him—known as Durte Dom was accused of sexual assault but had a very different experience of what happened next compared to Spacey (O' Connor and Hayock, 2022). All the Vlog Squad members were held accountable, but who were they held accountable by? The answer is you, the public, anyone who watches their videos. Most made apology videos and then took some time away from posting before making a return. Some lost sponsorship deals but have managed to recover. As Durte Dom was the accused, he got the worst of it from the people, but a lot of that vitriol came from his decision to not issue an apology. Although he has stopped posting on YouTube, he continued posting to TikTok. While he will probably struggle to get a brand deal, nothing is stopping him from continuing to be a content creator.

Although allegations are a serious topic, it is not the only thing that someone may seek redemption for. The next example experienced his fair share of a community turning their backs on him, which was especially tough as it was the community that he built his career on.

Imagine you stumble upon a gold mine, and after a few months of collecting that gold on a daily basis, a higher power comes and takes it away from you. For Jarvis Khattri, better known as Faze Jarvis, that is essentially what happened to him when he was banned from playing Fortnite.

Faze Jarvis is an English pro gamer and content creator who now resides in Los Angeles, California. He was born in England on November 11, 2001. His older brother, Frazier (better known as Faze Kay), started a YouTube channel back in January 2012 on which he uploaded footage of himself playing first-person shooter games (Kay, 2022). Faze Kay later joined the popular gaming organization known as Faze Clan in December 2013, almost two years after starting his channel. He even managed to become one of the directors of Faze Clan (ESportspedia, 2021).

Jarvis followed in the footsteps of his brother and started his career in content creation back in January 2014 on YouTube where he would upload footage of himself playing the popular game, *Call of Duty: Black Ops II* (Treyarch, 2012). It wouldn't be until mid-2018 when he would take his YouTube content creation seriously with the launch of the battle royale game *Fortnite* that took the world by storm (Epic Games, 2018). A battle royale game is a game that takes on a *Hunger Games*-like approach. You are dropped on a huge map and have to find weapons and gear to survive against the other players also dropped around the map. Last person standing wins. Fortnite added their own little twist on the game, by incorporating collecting materials and building, similar to *Minecraft* (one of the more popular games of today). Faze Jarvis was amazing at the game and would later join Faze Clan in April 2019 as a pro *Fortnite* player (Jarvis, 2022).

As the popularity of the game *Fortnite* continued to rise, demand increased for content surrounding the game, and gamer content creators saw this as an opportunity. These creators already had a few ways that they would get paid. Some of their income would come from tournament winnings, and they also got income from ad revenue, subscribers, and virtual items as a benefit of using whatever streaming platform they were currently on. Creating videos from their gameplay became increasingly popular for gamer content creators in addition to the other things they were already doing as this was another income stream for them. Video creation and streaming of the same game tended to be very different types of content creation.

Streams are live gameplay, so a content creator is giving a look into their domain. Not much has to go into it outside of raw gameplay, and the focus here is more on interaction with the viewers. Video creation from gameplay is a lot different, as videos are much shorter and typically have a theme (usually something really interesting or exciting) that is going to make viewers want to click and watch the video. *Fortnite* gameplay video concepts that typically did really well were videos of high kills because it showcases the creator's skill. After a while, video sharing platforms became oversaturated with these types of videos, and content creators had to pivot and come up with new, interesting ideas that would get a lot of views. Controversy has always been a way to drive views up, and it appeared that Faze Jarvis figured that out and decided to try to leverage it to create viral videos.

In the video gaming community, cheating is the most heinous crime that could be committed. However, because everyone wants to be good at video games, so many people will still cheat if they can potentially not get caught. This led to the creation of the aimbot, a hack that allows players to lock on to opponents to get kills in shooting games, eliminating the need for skill (Nelson, 2019). As much as cheating is frowned upon, that type of content has great potential for a lot of viewers, if you can deal with the angry comments.

What would make for a more controversial video than one the best *Fortnite* players in the world committing one of the worst crimes in gaming? In retrospect, it's easy to look back and say that nothing could be more controversial than a video about this topic, but I'm sure Jarvis didn't expect the backlash that would come when he decided to post a video of himself using aimbot on YouTube. A video that was supposed to be a funny joke did get him a lot of views, but it also got him hit with the most feared thing in gaming: the ban hammer (Webb, 2019). This one video got him banned from playing the game *Fortnite* and banned from Twitch, the platform that he streamed on. Some saw this punishment as a bit too harsh, while others saw it as the platforms using Jarvis to set an example.

Even the famous *Fortnite* streamer Ninja had an opinion on the matter: "There's a difference between a content creator who has millions of subscribers, hundreds of thousands of followers, who gets banned from what literally makes him money, and...some kid who is just a piece of sh*t who has absolutely zero following, has zero money that comes from *Fortnite*, from gaming, and hacks" (Chalk, 2019). His idea is that content creators face a lot pressure to make unique and entertaining videos on a pretty consistent basis.

No matter what your stance, everyone wondered what was going to come next for Faze Jarvis. He then dropped another video where he was in tears apologizing for what he had done (Jarvis, 2019). I'm sure he was hoping for the ban to be lifted, but unfortunately it wasn't. A realization did come out of releasing the apology video though, which at the time was the most-viewed video on his YouTube channel (now currently sitting at twenty-one million views).

This meant two things: People care about Faze Jarvis (along the way while he's been gaming and creating content, he managed to amass a large community of people who watch his content because of who he is and his personality), and people cared about the controversy. He gained millions of subscribers on YouTube, not just because of superior gameplay but because people liked him. So, Faze Jarvis made a slight pivot and continued his career as a content creator without the gaming footage. When it seemed clear that he would not be unbanned, Jarvis decided to embrace the controversy even more. About two months after his apology video, he posted a music video about him being banned for life, and that video currently sits as his most popular video of all time (currently at twenty-seven million views) (Jarvis, 2020). As he makes new videos outside of gaming, he will also every now and then create video that touches on his controversy, like when he fooled the world into thinking he was playing again while banned (Jarvis, 2020). Faze Jarvis's career hasn't ended; it has just merely evolved into something new, and he was given his shot at redemption.

Based on the stories we just examined about creators, we learn that this new digital age creates opportunity for redemption. If you wrong the people, that is who you answer to, provided you haven't been convicted of a crime. This is all because content creators control how they post, what gets posted, and when they post, and nothing stops them from posting even after they have done something questionable that people aren't fond of. As with the example of Durte Dom, his money may have slowed down, but he still has a career. Again, that doesn't mean that what they have done goes unnoticed or that it is okay; it just means that they can continue to post, and it's up to the people to decide whether or not they will continue to support.

I personally don't believe that people should be silenced. I don't believe that we should infringe on the freedoms that these platforms bring, but I do believe that people should boycott and not watch the content of people who do the unthinkable. On the other hand, with traditional media, you would likely lose your job in an attempt to save face and keep a positive image. Many content creators, including some of the creators previously mentioned, have been in situations where they have lost sponsorship deals, but they do have the ability to bounce back and the opportunity to try to win back the hearts of the people. Content creators and their communities have a unique relationship. Forgiveness comes easier here similar to how it's easier to forgive a family member, but that doesn't mean they should get away scot-free. Though companies and individuals have their opinions on what is forgivable, it should be up to the people as a whole.

PART THREE

BUILDING A NEW-WORLD MINDSET

CHAPTER SEVEN

HERE TODAY, GONE TOMORROW

"Going viral isn't random, magic, or luck. It's a science."

—JONAH BERGER

Many people may have different feelings regarding the quote above; however, I believe it to be true. I'm not saying that some people don't get lucky, like going viral without knowing the science, but that doesn't mean the science isn't the reason why they went viral. Most people who go viral don't necessarily plan for a viral experience. They just happen to stumble across it.

Let's take a step back.

What does it even mean to go viral?

A dictionary definition would read something like this: going viral is when a picture, video, or an article or written piece of work spreads across the internet like wildfire in a short amount of time. Viral videos have been around ever since the internet became a place for the mass public. Back in the day these viral videos were mostly spread via email chain.

One of the first viral videos to hit the web happened in 1997. Most people who spend time on the internet have seen this video or some variation of it at some point. The viral video goes by the name "badday.mpg," and people speculate as to whether this is the first video to go viral. The video looks like it is some sort of surveillance or security footage of an office and is of a man who appears to be angry and frustrated, sitting in a cubicle at what we believe is his job. It starts with the man pounding his keyboard as his frustration rises, and then a few seconds later, he picks it up and completely embodies Babe Ruth as he swings it into his monitor. This was back in the nineties, so picture this monitor as a gigantic, dinosaur-looking piece of technology that you may or may not have seen in your grandparents' basement. The video is then followed up by a terrified coworker peeking over the wall just in time to witness the frustrated man kick his monitor across the floor (Veix, 2010).

At the time that the video came out, social media didn't exist—no Facebook, no Instagram, no Twitter, no YouTube, no TikTok—so the video spread across the internet mostly via email. It is actually crazy to think that with no social media and no high-speed internet, this viral video managed to stay around for over twenty years, to the point where now variations and clips from the original video are floating around some of these social platforms. This video serves as a great example of a viral video as it contains some of the elements that make videos go viral: the surveillance-type footage aesthetic, a video clocking in at less than thirty seconds long, the contrast of someone blowing up in a calm setting, and things being destroyed. In fact, it came across so perfectly that conspiracy chatter started to arise about whether the video was fake or not. That is often still the case for certain viral videos today. In this case people started to question different aspects of the video, for example, the fact that the desktop appeared to not be plugged in (Veix, 2010).

It just so happens that with this video, the public was right. It was staged. Unlike other staged viral videos, it was not created with the intention of going viral. The main actor, Vinny Licciardi, didn't even know the video had gone viral at first. A coworker had to let him know that he saw him on television. It was actually originally intended to be some sort of an ad for the security footage. Vinny worked at a tech company based out of Colorado called Loronix, and contrary to popular belief, he wasn't a frustrated, overworked employee. In fact, his company was a tech startup and was typically regarded as a fun place to work. Loronix was developing DVR (digital video recorder) technology for security camera systems. DVR technology was not as advanced or common as it is today, so they needed sample footage to demonstrate to prospective customers. Vinny worked with his boss, Peter Jankowski, the chief technology officer of Loronix, to create these demo videos.

Peter took on the role of director while Licciardi was the sole actor. Their original video idea that was filmed was of Licciardi using an ATM and getting caught robbing the company's warehouse. Licciardi had the idea of being a frustrated employee in the video, and that's when Jankowski came up with the final idea.

> *"We had some computers that had died and monitors and keyboards that weren't working, so we basically set that up in a cubicle on a desk."*

—PETER JANKOWSKI (VEIX, 2010)

After shooting the video, they converted it to a MPEG-1 file so it would be compatible with Windows Media Player, the program that most people back in that time had to use in order to view videos on their computer. They put the video on CDs and handed them out at trade shows, and over the next year the video got passed around and began circulating through emails. Before long, Licciardi was a viral-video star.

With the advancement of the internet today, it is much easier for videos or content to be distributed across the web with all these different platforms. However, with how easy it is to produce and distribute this content, the internet is oversaturated with videos that would have had viral potential back in the day.

That brings us to the next relevant creator, Adin Ross. Although going viral is more difficult to do today, he has managed to do it continuously in such a short span of time.

How?

Well, I think we should start off with who Adin Ross is.

Adin Ross started off as a Twitch streamer who would stream himself playing video games, specifically *NBA 2K* (Ross, 2022). He has since evolved into a complete entertainer, and with his followings rising on all the platforms, he has become a content creator that can't be limited to one social platform. He has well over a million followers and subscribers on four different platforms: Twitch, YouTube, TikTok, and Instagram (Skelton, 2021).

In the video game-streaming world, different communities surround different types of games. One of the communities is the *2K* community, which is a bit smaller than some others but where Adin Ross got his start. Looking back at his story, it would seem as though he always had a master game plan, but a bit of luck was involved. He became friends with Bronny James, the son of Lebron James, and the two would play *2K* together. Adin started getting his first bit of viral attention by placing wagers on games that he and Bronny would play against other streamers. Another big moment came when Ross got the opportunity to talk to Lebron briefly, and a clip of the conversation went viral (Skelton, 2021).

After this, with his now bigger following, Adin moved on to hosting "e-dates," a segment where he would set up a bunch of streamers to compete for a popular Instagram model. It was essentially an entertaining virtual dating game show on his Twitch stream (Adin Ross, 2022). These did well and showed that he could entertain and create content outside of gaming, something which is very difficult to do in a live setting.

Then the game *Grand Theft Auto V* made a comeback right at the end of 2019 and beginning of 2020. With role-playing servers being launched, most streamers jumped on this opportunity to play what would prove to be a very entertaining game, and Adin was one of the streamers who got on board (Rockstar North, 2013). At the same time, COVID-19 forced everyone to be inside their homes. Everyone, including famous rappers and athletes who now had a ton of extra free time and nothing to do, decided to game as well. Adin made a lot of connections with these people while playing the game, and he also had a lot of viral moments during this time by trolling these celebrities with "sus" (suspect) jokes—jokes that lead you to question his preferences all in attempt to get a reaction out of these celebrities.

In early 2021, Adin Ross got the opportunity to move into a content-creator house in Los Angeles, and with all the buzz surrounding him, it was the perfect time to catapult his career. He used his newfound fame to collaborate with other creators, rappers, and A-list celebrities, and when he collaborated, he would be sure to do some "sus" joke trolling, which would lead to funny reactions that would, in turn, make him more viral. As he continued to go viral, more people kept reaching out to him, and his collaborations grew (they saw him as a free ticket to clout), which started what would seem like a never-ending cycle.

Going viral in this new digital age is priceless. People will pay top dollar for it, and learning the way to go viral could prove to be very beneficial no matter what you do today as a career. Viral moments lead to more eyeballs or followers on you, which could then be leveraged and turned into anything, such as sales, votes, or followers on another platform. The key here is paying attention to the trends as well as the audience.

> *"Every time you think of the word 'algorithm,' replace it with 'audience.'"*

—JIMMY DONALDSON (POONIA, 2022)

That's coming from the fastest-growing YouTuber on the planet. He has managed to figure out a process that works well and is successful, all through paying attention to the trends and audience.

CHAPTER EIGHT

THE INFAMOUS INFLUENCER

"The villain is the character that the people remember."

—UDO KIER

In this new world, being loved is just as valuable as being hated, but what's more important is how much you are hated or how much you are loved. In this new space we are venturing into, emotions online drive viewership, and views push revenue into influencers' pockets. Countless influencers embrace the villain persona. This is something that I believe derives from sports. In sports, the bad guys always draw in viewers. For example, back in the day, the NBA had the "Bad Boy" Detroit Pistons, who were known for being the bullies around the NBA, and people would watch in hopes of seeing the bullies get bested. They had guys like Bill Laimbeer on their team who would stop you by hurting you, like what they did to Jordan by hitting him out of the air in hopes it would stop him from going to the basket (Golliver, 2020).

The real roadmap to follow came from a man playing a nonteam sport, the boxing legend Floyd Mayweather. Floyd was born on February 24, 1977, in Grand Rapids, Michigan, and he comes from a family of professional boxers, from his father to his two uncles on his father's side of the family. He was destined to become a boxer. Floyd had a tough upbringing, as his mother was addicted to drugs, and his father was a drug dealer. His father was a part of his early life, though, before he was incarcerated. His father would take him to the gym early in the day to box and then would sell drugs at night. When his father went to jail, Floyd put all his focus into the sport of boxing, even taking it as far as dropping out of high school to put his full focus there. He then he went on to an amateur boxing career that ended with a record of eighty-four wins and eight loses.

What Mayweather was most known for outside of his defensive fighting style was his antics outside the ring (and actually sometimes in the ring). He was well known for taunting his opponents, and when he went pro, he made sure to put more focus on this. Because he was smart, he knew this would sell the fight, and he was good at it (Biography.com Editors, 2019).

I know people have done this before and after Floyd in boxing as well as in other sports, but I believe he did it the best. His villain persona mixed with his undefeated record made him the perfect person for people to hate. So many people wanted to see him lose, and fight after fight, millions of people would pay for the pay-per-view to watch (Dawson, 2020). He even bought out his promotion contract and started his own promotion company, because he knew how to sell the fights and didn't need to get ripped off by another company. Win after win, he drew more people in, and more people were invested in the villain that was Floyd "Money" Mayweather. Because of that and some other very smart business moves, by the end of his career as a boxer he would be the highest-grossing fighter of all time (Parker, 2022).

This brings us into another person who takes on the supervillain role as a content creator, Jake Paul. Jake is one out of the two Paul brothers. He was born on January 17, 1997, and grew up in one of the suburbs of Cleveland, Ohio, with both his parents and his older brother. One year for Christmas when he was ten years old, he and his brother got a video camera from their father. They then became a brotherly duo filming funny videos all around their house, which they posted to YouTube, something that got them some mild popularity at school. A few years later when Jake was be a sophomore in high school, he decided to join the wrestling team, which he was very passionate about and took so seriously that he slowed down video making with his big brother. But things changed when an app called Vine came out in 2012. He and his brother managed to gain a ton of popularity relatively quickly on the platform through their short video posts (Leskin and Greenspan, 2020).

Jake attributed their fast success to their personalities and luck.

"We didn't care what people thought. We were the loud brothers from Cleveland, kind of crazy, and that made us relatable. We were in the right place at the right time, and we were making more money than our parents before we knew it."

—JAKE PAUL (LESKIN AND GREENSPAN, 2020)

Soon after, Jake decided to do his senior year of high school online and move to Los Angeles with his brother to continue to push their new social media careers forward, and that's exactly what happened. They were getting roles in small films and brand deals, and Jake even managed to go mainstream when he landed the role of the main character on a Disney Channel show called *Bizaardvark* (Mendoza, 2016). Vine was shut down in 2017, and at the time Jake had already racked up 5.3 million followers and almost two billion video views. He couldn't just let his social media career end there, so he transitioned to YouTube.

Jake's YouTube channel was comprised of pranks, vlogs, drama, and controversy (Paul, 2022). From faking relationships and even a marriage, to vandalism and scamming his young fans, he was a very polarizing figure. It seemed like controversy just followed him around and led to Disney Channel letting him go as that did not go with their brand. That did not stop him, though, as he managed to amass over twenty million subscribers on YouTube over the years (Chen, 2018).

That is why some people found it strange when he decided that he wanted to become a boxer and venture into the world of fighting. He got his first opportunity being the undercard for his brother's celebrity boxing match.

> "When I moved to Los Angeles to act and make content, it was a great opportunity, but I sort of left sports behind. When I got the opportunity to box when I was twenty-one, it brought back the old Jake Paul. It brought back the competitive, athletic Jake Paul. I just fell in love with it. When I won my first boxing match, it was one of the best feelings I've ever experienced in my life."—Jake Paul (Chen, 2018)

The world wouldn't know it yet, but Jake's supervillain persona was perfect for the sport of boxing in terms of selling pay-per-view. This is precisely because people want to see the villain get beat, and in the sport, it's possible to witness a villain fall and get knocked out. He proved it when he fought against Ben Askren in early 2021 and generated $65 million from 1.3 million pays-per-view, making him one of the highest-paid fighters that year (Jake Paul, 2021). People who once questioned his decision to get into boxing were now forced to pay attention.

While exploring this new digital world, it is safe to say that as time has passed, content creators are learning more and more, and they have managed to develop into some of the best marketers of our time. Some have created these personas that sell themselves; some are so good it has become hard to tell whether they are wearing a mask or if that is truly who they are, and that is exactly what gets people to tune in. Some of these content creators have taken it so far that we hate them and hate them enough to tune in to whatever they are doing and hope they lose or fail. These creators live for the hate. It leads to more views which, in turn, leads to more money.

> *"I'm polarizing. I rub a lot of people the wrong way. There's the other side who have really embraced me. I'm not mad at it. I am who I am. My friends are my friends, and like Drake said, 'I don't need new friends.'"—Jake Paul (Rawden, 2021)*

I do believe that people are catching on to the content creators and to some of these personas being a facade, but that doesn't stop them hating as being a part of the community that hates similar things is fun; it is like having your own real-life supervillain. Content creators who took this path knew this, and that is why they are considered marketing geniuses. It doesn't even need to be limited to the supervillain persona, but having a persona that people can latch on to is crucial. I would even go as far as saying that some creators have taken the opposite approach like YouTuber MrBeast who gives away so much to the less fortunate in his videos and could be regarded as a hero. That path in itself has its own challenges. Though with people always searching for a slip up, infamous influencers don't have to worry because for them, it's just another controversy.

CHAPTER NINE

THE SUPERFAN

*"I'm very thankful to all my fans for their
constant love and support. I am what I am
because of their unconditional love."*

—MAHESH BABU

Throughout time, celebrities have had fans; that's part of
the reason they are considered celebrities. With the rise
of social media and this new digital age we have entered,
a new wave of celebrities was created. The influencers
and creators are modern-day celebrities, known and loved
by many, and they get stopped in public for pictures and
autographs. Even though many of them may not feel like
they are celebrities because of their come up, there is
no doubt that they indeed are. These new-age celebrities
have bred a new type of fan: the superfan. Superfans are
basically a type of die-hard fan, the type of supporters
who look out for whatever new projects or products their
favorite creators make and who are going to support it no
matter what. These superfans came out of people developing parasocial relationships with their favorite celebrities.

"Parasocial relationships are one-sided relationships, where one person extends emotional energy, interest and time, and the other party, the persona, is completely unaware of the other's existence. Parasocial relationships are most common with celebrities, organizations (such as sports teams) or television stars…In the past, parasocial relationships occurred predominantly with television personas. Now, these relationships also occur between individuals and their favorite bloggers, social media users, and gamers."—National Register of Health Service Psychologist (Bennett, Rossmeisl and Turner et al., 2022)

With these influencers and content creators sharing content so frequently and also giving many glimpses into their personal lives, parasocial relationships became elevated. When it comes to your favorite actor or actress, you see them whenever their next movie comes out; when it comes to your favorite content creator, you see them almost daily through their constant posts and more intimate content. It is like you are truly getting to meet your favorite content creators, like you truly knew them.

These superfans have changed the game. Not only will they tune in and bring the views, but they will also support their favorite creators as if they were their own family because that's how they actually feel most of the time.

In the last chapter we talked about the Paul brothers but mostly focused on the one with the supervillain persona. Here we're going to be looking at Logan Paul and something amazing that he was able to create, the Maverick Club (Maverick Club, 2022).

So, as we all know, Logan isn't any stranger to controversy, and this controversy has sometimes been worse than others. Sometimes, because of things he had done, his videos weren't being monetized on YouTube, and other times brands did not want to associate with him because of things he had done—most notably the fiasco where he vlogged being in a forest in Japan that was known for suicide and recorded someone who had just taken their own life. After this, he apologized and took a long break from posting (Abramovitch, 2018). Often times, depending on how bad a prank video is or if there is profanity in a video, it could become demonetized. This is a problem for content creators because that is how they get paid, and for some it is how they are able to fund more videos.

Logan Paul created the Maverick Club essentially as a platform that only had his content. This meant that there were no restrictions, and he could post whatever he wanted. The platform also runs on a subscription-based model where Logan charges $19.95 per month for access to the platform (Maverick Club, 2022).

> "Paul's Maverick Club is a genius business move on various fronts. Not only does the subscription service help Paul build an even stronger fanbase, but it gives him his die-hard fans' information. Once a user is signed up, the Maverick Club now has access to the names, ages, and contact details of his most loyal fans. This information is extremely powerful as a marketing tool

to send out mass emails for future merchandise drops, boxing matches, and other possible ways to get fans to re-engage."—Brendon Cox (Cox, 2021)

This is something that a lot of other creators are mimicking right now in an attempt to try to leverage their superfans.

Another example of a creator with superfans is the creator group the Nelk Boys. They are a group of YouTubers with over 6.65 million subscribers on their channel, yet the platform isn't the biggest fan of them (Nelk, 2022). Their videos can be quite edgy, and that leads to most of their videos being demonetized, like when they pretended to be tow truck drivers and actually had cops show up (NELK, 2021). They do, however, have a very strong fan base which they leverage through merchandise drops. The have created a brand called Full Send, and they sell clothes and other small pieces of merchandise (Full Send, 2022). They do a few drops each year—whatever merchandise is in the drop will never be released again once sold out—and they sell out in minutes every time.

Another content creator group called Yes Theory basically does a similar thing as the Nelk Boys. They too have created their own brand by the name of Seek Discomfort (Seek Discomfort, 2022). They also do merchandise drops in the same way; once things get sold out, they will not be brought back out, and they manage to sell out pretty quickly as well.

These brands they have created are an extension of their YouTube channels. The Nelk Boys' channel is all about partying or pranking and having a good time like college glory days, and Full Send, which basically means to go hard at a party, is very on brand. The Yes Theory channel is actually about saying yes to things that take you out of your comfort zone, and the name of the brand is Seek Discomfort. The fact that these brands are mere extensions is why superfans are so quick to support them, and the additional support comes from fans actually liking the products being offered.

The concept of superfans is very interesting, like brand loyalty but to a whole new level. If someone had brand loyalty to Apple, when they needed a new phone they would get an iPhone. With the concept of superfans, they would buy out the entire store. Superfans are interesting because they are almost a type of business model, and certain products only work if the content creator has superfans. Some content creators even rely on superfans in order to fund their operations because their content may be too edgy for brands to get behind as we saw with the Nelk Boys.

Essentially content creators with superfans are in a very unique place as they have managed to gain a following of people who will listen and, better yet, open their wallets for them and who will stay after controversy and support them like they are family. This is an opportunity for businesses, brands, events, etc. Working with these content creators would be beneficial for them as they would gain the support from their superfans as well. They follow the advice from these creators, and there is real trust there that developed from merely watching their content. As we are learning more and more about this new digital world, we see that people have taken notice of this and have started to take it to the next level, something that we will discuss in a later chapter. The fact of the matter is that with the digital world growing larger and larger, this concept will continue to evolve, and I could foresee super-fans paying for more than T-shirts and actually investing in these creators, especially with the rise of Web3.

PART FOUR

HOW TO SUCCEED IN THIS NEW WORLD

CHAPTER TEN

KIDS WILL LEAD
THE NEW WORLD

*"Parents are the ultimate role models for children.
Every word, movement, and action has an
effect. No other person or outside force has a
greater influence on a child than the parent."*

—BOB KEESHAN

I am not a therapist or psychologist by any means. I
am not the parent whisperer or some sort of guru. I am
merely a son and bystander, offering my thoughts based
on my experiences and what I have witnessed.

Growing up I heard a lot of the statement, "Put down
the game and pick up a book." I loved video games when
I was growing up, and honestly I still do. It wasn't just
video games though; it was technology in general. I was
amazed at all of the products that were coming out and
their evolution in such a short time. I also had a passion
for creating from a young age. I would build things out
of anything I could get my hands on—shoe boxes, wire

hangers, anything. At one point I combined these passions and created content before I even knew what content was. I even posted some of this content to YouTube. The videos are private now, but I haven't deleted them, and I don't think I ever will. I posted my first video to YouTube when I was thirteen years old, with no help or prompting from anyone, which I think is special. At the time content creation wasn't a viable career path in the eyes of adults, and my parents viewed it as me playing around with games. What if instead of prompting me to "get off the game," they had encouraged me to pursue and learn more about my interests? This is not to bash my parents in any way but to show that it was a different time.

Today we live in a world where my baby cousin knew how to navigate YouTube on my phone before she could speak in full sentences.

While doing extensive research and interviewing, I managed to get one of the most informative interviews with a smart eleven-year-old in 2021. I wanted to explore some of the differences between a child from today and compare it to my childhood, and I learned so much. When I was eleven, I didn't have my own device that could access the internet yet, only the family desktop computer, but the child I interviewed had a laptop. I spent my free time playing video games but mostly watching television; he spends most of his free time on his laptop, playing online games or watching YouTube. I also asked who his favorite celebrity was, and he said Flamingo, a popular YouTube content creator; when I was eleven, I believe my favorite celebrity was Chris Paul, an NBA player (Flamingo, 2022).

The most interesting answer that I got from him came after I asked what his dream job was:

"I want to be a YouTube influencer or a professional swimmer."

—SMART ELEVEN YEAR OLD

I found that to be interesting because back when I was eleven, being a YouTuber wasn't considered a valid career path. That wasn't something that someone would aspire to do for a career, but that just goes to show how much the world has changed in such a short time.

I bring that up to explain that parents need not be worried if that is a career path that is now desired by their child. So many people have created a career out of YouTube. Don't shut down their idea but rather help them to explore how they could potentially achieve their goal.

That brings us to one of the youngest content creators out right now, a six-year-old who goes by the name Rowdy-Rogan. He streams and makes gaming content with his father that they post to YouTube. Aside from his father being into video games and wanting his son to be like him, Rogan also just had a talent and understanding for these games from a very young age. His ability and drive at such a young age has amassed hundreds of thousands of followers and subscribers across multiple platforms (RowdyRogan, 2022). The support given to him by his parents plays a major role in his success. He is only six years old, so he is limited in how much he can do on his own at this point. Making exceptions to allow him to play hours of video games to practice and get better is what is going to have him stand out against other kids whose parents don't allow them to play that many games. His parents are not unique with this thinking. Times are changing, and some parents are starting to recognize the benefits to gaming.

Nearly 63 percent of parents said that video games have a positive impact, across all age groups, though that wanes a bit as the kids get older—only because social media takes over as the "most influential" content for kids ages fourteen to seventeen. Parents also like the fact that gamer kids have good logical thinking skills (62.9 percent), and, well, they're happier (60.5 percent) (Griffith, 2021).

Another topic to consider for parents is how to discipline their kids in this new digital world. I'm no parenting guru, but it seems like a bad idea to punish a kid interested in pursuing a career in gaming by taking away their video games or to punish a content creator by taking away their camera. This brings us to another couple of content creators, the D'amelio sisters, Charli and Dixie, two famous TikTokers. Charlie actually happens to be the most-followed person on TikTok, amassing over one hundred million followers. They were on the *Views* podcast when they were asked what it's like when they get in trouble, a normal question since at the time they were sixteen years old and nineteen years old. They responded that before TikTok when they would get in trouble, their parents would take their phones way, but now that is not possible since TikTok is their work (VIEWS, 2021). It's probably not the best idea to take away phones when that's what pays the bills.

Times are different now; we are all having to learn about the new digital world and adjust, but parents have to adjust for themselves as well as for their kids. The most important thing to remember is to remain opened minded and to not be afraid to ask questions to get more information. It is going to be increasingly important for parents to recognize the differences in when they grew up and today. Aside from the cultural differences, career options and ways to be successful are different now too. Kids are the future, and it is imperative that we monitor what they're doing but equally important that we watch what they gravitate to. In the years to come, they will be responsible for carrying on society, and they will influence the markets. We have to be careful what stance we are taking with them: remember to guide but not hinder.

I mentioned a few stories above about some kids being the ones bringing in the fortune to the family at a young age. That will be the reality sometimes in this new world, and learning how to deal with it is important. Children are going to make mistakes, but it can be hard to correct them if they are the breadwinner. Kids will still be kids, and no matter how much early success or fame may come, they will still need guidance, which can only come from an adult who loves them. It will be tough to navigate in the beginning, but once you keep an open mind it will get easier. Kids are ultimately going to be a lot of the leaders in the new digital world, and they will gravitate easily to the new technology and learn it a lot faster. It is important to put them in positions to be successful. Allow them to explore their interests and to reach as far as they can.

CHAPTER ELEVEN

PIVOT OR DIE

"When you're finished changing, you're finished."

—BEN FRANKLIN

One thing that has been consistent for companies throughout the years is pivot or die. That is a play on words after the popular business catchphrase "change or die," which was coined by business author Alan Deutschman (Boomer, 2016). At one point I remember being excited for the weekend to take trips to Blockbuster; then I remember one weekend we didn't go to Blockbuster but instead we got a DVD in the mail—the start of Netflix.

Think about it: Forty years ago, the internet didn't exist, and now about 60 percent of the entire world's population has access to it (Johnson, 2021). About twenty years ago you probably didn't have an email; now practically everyone does. The world is constantly changing and forcing us to adapt to survive. That statement has also been true for businesses over the years. In fact, the initial rise of the internet was the death of a few companies who refused to adapt to the changing times.

With the rise of social media, traditional media companies and big studios have the most to worry about at this time. Right now, they should be considering smart ways to break into the world of media on the internet. Some companies have already caught on and started making changes, for example Disney creating Disney+, their own streaming platform (Tapp, 2022). Companies like Netflix, even though not a traditional media company, also need to keep adapting; one way that they have started is by hiring some famous content creators to star in some of their films.

Certain social platforms have been very active during the rise of socials, and they have stayed vigilant in order to continue to put their platform in the position to succeed. That is exactly what Instagram is doing by announcing that they will be launching new Instagram creator monetization features. They announced that there would be three new features.

The first creator monetization feature is creator shops.

"[For creators] who want to sell your own products and merchandise, we're going to make it easier to add an existing shop or open a new shop on your personal Instagram profile."

—MARK ZUCKERBERG (DEMEKU, 2021)

Their overall goal is to help content creators, especially the smaller ones, make more money, so they are aiming to help users purchase directly from creators. Instead of figuring out and creating their own online store, creators can use the platform and infrastructure that is there already. For full-time creators, time is a huge constraint, so not having to worry about other time-consuming things will be helpful.

The next feature is affiliate commerce.

"The affiliate tool [will] allow you to discover new products that are available on Checkout and share them with your followers to earn a commission for the purchases you drive, all within Instagram."

—MARK ZUCKERBERG (DEMEKU, 2021)

This plays into what we discussed in the last chapter. This will be an easy way for content creators to be supported by their communities.

The last feature is a branded content marketplace.

"We should be able to help brands find creators that are uniquely aligned with the work they're trying to do and vice versa. If we can help with matchmaking, we can help drive more dollars

*to the smaller creators who can do amazing
work for brands."*

—ADAM MOSSERI, HEAD OF
INSTAGRAM (DEMEKU, 2021)

This last feature comes from the idea of TikTok's Creator Marketplace. Instagram is aiming to step in and join the party.

> *"The new tools are still in the works but could dramatically shift the way influencers monetize their following on the platform. Many of Instagram's top stars already run online shops and form partnerships with brands. But for now, those deals largely happen off-platform so it can be difficult for less well-known personalities to make money. Bringing these kinds of tools into Instagram could make it easier to land deals. But it would also give Instagram more control over its creator ecosystem and incentivize influencers to spend more time on Instagram than other platforms."—Karissa Bell (Bell, 2021)*

This isn't the first time Instagram made some pivots to stay ahead. After Snapchat was launched and saw some major success in the beginning, Instagram and Facebook decided to join in on the party by adding twenty-four-hour story features to their platform which quickly grew in popularity. Also, after TikTok was launched and grew in popularity, Instagram added reels, an opportunity for content creators to drop short-form content videos on Instagram.

When Facebook realized how big the gaming industry and streaming were, they jumped in on that party by creating Facebook gaming. With this platform, they made sure to add things that the other main platforms in the space like Twitch did not have. They did a way better job than any other platform with on-site promoting: after watching videos on Facebook, viewers can see a random stream pop up, which helps smaller streamers grow.

YouTube also has shorts, which was YouTube's way of getting in on the short-form content game. Everyone sees the opportunity for it with the massive success of TikTok, and no one is trying to get left behind. They added it to mobile in an attempt to get more people to use it. This feature has experienced a good bit of traction, as creators have found that it can lead to more people finding their page and are really taking advantage of it.

I also see an opportunity for financial institutions to step in and get a piece of the pie if they're smart about it. These institutions should find a way to get money into the hands of creators like they do with entrepreneurs with small business loans and other programs. They need to do this soon while they still can because as things keep evolving in the digital world, fewer creators are going to need outside funding.

The same way that we have to learn this new digital world and adjust to be able to find new opportunities, companies and businesses need to do the same or else they will not survive. The concept of pivoting for companies isn't new. Big companies that are still around today that have been around for a long time already have some experience in pivoting; they had to do it, or they wouldn't be around anymore. Businesses and companies also have countless examples of entities that went under because they did not pivot or didn't do so fast enough, like Blockbuster not realizing things were moving online and that people would prefer convenience. That is what this book and chapter specifically are going to prove to be so important. Even social media platforms need to continue to grow, adapt, and pivot with the new world that they brought about, because if they don't, competitors will rise past them. As long as companies keep that in mind and adapt to the changing environment, they will be fine. The problem is that things are changing so fast they have to be on board to follow and pivot now if they haven't already. Companies need to stay on top of things, listen to audiences, pay attention to the data, and not be afraid to take calculated risks.

CHAPTER TWELVE

THE COVID ACCELERATOR

"To be successful you have to be lucky, or a little mad, or very talented, or find yourself in a rapid growth field."

—EDWARD DE BONO

March 11, 2020, was the day that changed life as we all knew it.

I was sitting on the balcony of my hotel room admiring the beautiful view in Cancun, Mexico. This was spring break of my senior year of college, and I was on vacation with my friends, but for this brief moment I was enjoying some alone time, thinking about what was next, thoughts that were never foreign to me. I had just finished having an eventful day on the beach, and I thought what else would be a better way to recharge than pondering thoughts of the future? I mean, after all, it was senior year, but no one could have predicted what would come next.

Being the huge basketball fan that I am, I had set an alarm for when the NBA game was set to start. The ringing started, and I immediately went inside the room and flipped to the correct channel. Something was a bit weird about this game though, as it was taking unusually long to start. Then the halftime performers came out and performed, and that's when I realized something had to be wrong. After the performance, my thoughts that something was wrong would be confirmed. A few of the players set to play that night had contracted COVID-19, and the game was postponed, the start to what some would call a nightmare.

I knew about COVID-19, the world knew about it already, but watching the NBA get cancelled on live television made it feel so much more real. That wouldn't be the end to that day, as I flipped to the news and apparently Dr .Fauci was also called to testify about the coronavirus before the House Committee on Oversight and Reform on that day as well. That's where he spoke words that would haunt me.

"It's going to get worse."

—DR. FAUCI (WAMSLEY, 2021)

Not knowing what else to expect, I remained glued to the television screen for a while. I wasn't supposed to meet up with friends for dinner until 10 p.m. so I had some time, as it was around 8:30 p.m. when the NBA game was postponed.

How much more could happen within that hour and a half? The answer is a lot more than I would've expected. The president addressed the public and issued a travel ban on Europe. This was even scarier for me because I was out of the country at the time and had a few more days before my flight home. I was nervous that all travel would be banned, and I might be stuck in Mexico. Minutes after watching the president's address, word comes out that Tom Hanks and his wife, Rita Wilson, had contracted the virus (Gonzalez, 2020).

Fast forward a couple weeks, and I'm home with my family trapped in the house, finishing my last bit of classes for senior year virtually. When classes were done, I had a ton of free time on my hands just like many other people, so I turned to video games. One game in particular really caught my attention, *Call of Duty: Warzone*, a battle royale-type video game that I could play with my friends online (Raven Software and Infinity Ward, 2020). During this time, I got so caught up in the game that I started watching streamers play the game as well. With nothing to do and everyone stuck in their houses, people needed more content to consume, which led to a very prosperous time for content creators.

One of the creators who saw major growth thanks to everyone being at home was Kris Lamberson, better known as Faze Swagg. He actually joined the Faze Clan, the gaming organization mentioned back in chapter four, in April 2020, about a month after *Call of Duty: Warzone* had been released (Faze Clan, 2022). Like most other streamers, he got to the game as soon as it launched, and early on, it was clear that he was one of the better players. He gained so much popularity during that time that rappers and NBA players made guest appearances on his stream to play *Warzone* with him. The game also led to him getting into esports, competing in multiple *Call of Duty: Warzone* tournaments, and even taking home the victory in some of the biggest tournaments. His skill and entertaining reactions led to popular streamers wanting to play with him, which propelled him into much-deserved stardom. Although COVID-19 accelerated his growth, that doesn't mean he wasn't already heading there. He was already a reputable gaming content creator who would post videos to YouTube since 2013, as well as a decently sized streamer (Stedman, 2021). That being said, the pandemic did increase gaming livestream audiences by 70 percent (Stream Hatchet, 2020).

Another gaming streamer that gained a lot of popularity during the time of COVID-19 was a man named Jeremy Wang, better known as Disguised Toast. His accelerated success came from the fast-growing popularity of a game called *Among Us*, a fun detective-type game to play with friends. The goal of the game is to figure out who the imposters are before they manage to kill all the crewmates. Disguised Toast grew a lot during this time due to his ability to create very entertaining YouTube videos out of his gameplay. He would somehow be great at figuring out who the imposter was with his superb detective skills as well as coming up with great alibis when he was the imposter, and it all made for great videos. He would post a new *Among Us* video every day for eight months— just what people needed during this time, more content.

Because of this, he saw an impressive amount of growth which led to him being on the *Forbes* 30 under 30 list in the gamers category (Moore, 2021, Forbes, 2022).

Gamers weren't the only creators that saw growth during this time, though. Content of all types saw increases during the pandemic. Even professionals were getting in on the action, creating videos out of whatever they do, thus making themselves content creators as well. Dr. Daniel DeLucchi was one of these professionals, a chiropractor who would post videos of his work, and during the year of 2020, he grew to 443,000 followers on TikTok (Wong, 2020).

This pandemic that left many of us scrambling and confused also gave us insight as to what is to come in this new digital world. Content creators' growth was accelerated as well as the growth of some of these social media platforms. The pandemic gave us the snapshot that we needed to see; people want more content, and as time goes on, our appetite for content increases. So much opportunity already exists, and some people have taken advantage of it by jumpstarting or propelling their content-creation careers. Some may think that the opportunity has come and gone, but I am here to assure you that it has not. More and more people are still coming to these platforms, like TikTok, and are still growing on a daily basis. The internet definitely has room for you as a content creator.

CHAPTER THIRTEEN

INFLUENCER 2.0

"I'm not a businessman, I'm a business, man."

—JAY-Z

We are officially entering a time period that I refer to as Influencer 2.0 or the time of the unorthodox creator. We may have gotten here a bit faster due to COVID-19, but nonetheless we were heading in this direction regardless. Some people might not know what I'm talking about but not to worry. I will explain.

We already got to the point where companies are willing to use social media influencers to promote their products. Depending on an influencer's followers and engagement, they could warrant different amounts in these brand deals. I call the time period that I just described Influencer 1.0, a very exploratory time. Companies were seeing whether it was worth it and eventually realized that it was. Big brands and companies started to reevaluate their marketing budgets. Imagine that—a piece of the budget being designated for influencers. Now what is Influencer 2.0? Do you remember the hypothetical from the beginning of the introduction? The kid who is a streamer who is so big that he has more viewers than a full football stadium, and the streamer also had his own cereal brand. Well, what if I told you that things similar to that hypothetical have already begun to happen?

The biggest streamers have gotten to that point. They may not all be on that level yet, but some are and not just in terms of the viewers, but in terms of the brands. Influencer 2.0 is the time where social media influencers have figured out the game and are now creating their own brands. Why promote someone else's makeup brand when you could create your own? The picture that was painted in the beginning of the book's introduction may have seemed a bit farfetched, but when looking at the reality of where we are and what is already going on, it seems that we are actually in this reality. We are currently in the early stages of it. In fact, one content creator that we have discussed earlier in the book is already living the hypothetical from the introduction but a different version of it.

That content creator is Jimmy Donaldson, better known by his YouTube name MrBeast. While he's not a livestreamer, he is a content creator that makes YouTube videos, and it's safe to say that his views have surpassed what we may see as normal (MrBeast, 2022). Just to put into perspective the type of views he gets, here's this example.

Popular streaming service Netflix released an original show called *Squid Game*, and they reported the show pulled in 111 million viewers in the first month, a Netflix record. Jimmy created his version of *Squid Game* in real life for a YouTube video and pulled in 138 million views in the first week (Dong-Hyuk, 2021, Alford, 2021). Now that we see that he reaches pretty crazy numbers, we can just imagine how many companies would love to get their products in his hands so that he could promote it to his millions of viewers. Like most content creators when they get a decent following, he sells merch branded with the logo from his YouTube channel. Now Jimmy still takes brand deals, but he decided to do something bigger, create something that could potentially outlive his content creation if he ever decides to stop. he decided to create a restaurant, MrBeast Burger.

He did this by means of a "ghost kitchen." What does this mean? A ghost kitchen is essentially a restaurant outsourcing the cooking of their food to other restaurants, and it is in my opinion the easiest way for a busy content creator to start a restaurant and still focus mostly on content. Jimmy actually partnered with another company for this endeavor, a company called Virtual Dining Concepts. The startup specializes in linking preexisting restaurant kitchens with new virtual restaurant concepts. This means that an up-and-running restaurant kitchen can make food for another brand while maintaining its own establishment (Watson, 2021).

According to the MrBeast Burger website, it is available in forty-nine out of the fifty states, as well as Washington, DC; the only state it is not available in is Montana. This was only able to happen so fast because of the ghost kitchen system (MrBeast Burger, 2022).

Instead of spending money on advertising, MrBeast simply tweeted about the launch and made a viral YouTube video. There was no need to focus on traditional marketing, especially in this new age that we're in. The appropriate decision is to focus on social media. But MrBeast Burger already held an advantage, and that was MrBeast. There was no need to pay for Instagram ads or Facebook ads. Across all platforms MrBeast has easily over one hundred million supporters. He built a following so large that paying for marketing is redundant, as anything he posts goes viral, and that was the case for his posts about MrBeast Burger.

To understand this, go back to 2019, when a decidedly nonvirtual concept, Popeyes Louisiana Kitchen, introduced its chicken sandwich. A single tweet about that sandwich led to an explosion in interest and one of the best quarters a single restaurant chain has ever seen. In two years, that restaurant chain has generated $1.3 billion in sales (Maze, 2021).

> *"You have social media now. In the old days the only way to get food out to the people was to have a great restaurant and to make great food. People would talk about it. They'd put an ad in the paper, radio or TV."—Mark Wasilefsky (Wasilefsky, 2021)*

Basically, the thought process here is how good does the food need to be if you could potentially make it go viral? That doesn't mean that you shouldn't care about taste at all, as that is what will make people continue to come back, but being viral can easily get millions in the door to start. That thought process extends far beyond just the restaurant and food space. It bleeds into all industries today. With content creators starting to realize this, we will start to see creators and influencers venturing into new spaces and building their own brands, which leads us back to the Nelk Boys.

The popular YouTube group started their own alcoholic seltzer brand, Happy Dad (Happy Dad, 2022). It follows their college-kid, party-themed brand. I'm sure there were many alcoholic beverage brands that reached out to them to potentially have them promote their drinks, but the Nelk Boys realized how much influence they had and took it upon themselves to start their own brand from scratch (NELK, 2021). From the taste testing, branding, strategizing, planning, and getting the product in physical stores, they took the more traditional approach as opposed to what MrBeast did.

This concept of influence bringing in your first customers travels far beyond just the food and drink industries. The next example shows that if you choose the right industry that fits your personal brand that you've been building for years, then it could be more successful than you would've ever imagined. This is exactly what happened with Kylie Jenner and her celebrity beauty brand Kylie's Cosmetics (Kylie Cosmetics, 2022).

Kylie stayed within her realm as she promoted, marketed, and built her brand. This was what she was already going to do with another beauty brand minus the building part, so why not create her own and let her labor bring forth more rewards? She did that, and in 2019 sold 51 percent of her cosmetics and skin care brand to Coty for $600 million, a price tag that values the enterprise at $1.2 billion (Friedman, 2019).

What's important to get out of these different stories is that all the parties involved are content creators and social media influencers who make and post content online. They managed to use the popularity and influence they gained to start a virtual restaurant, a traditional seltzer brand, and a beauty brand—all of which are bringing in quite a bit of money—not because of the quality of the food or the convenience or because they have the best products but simply because of the name on it. Similar to what we discussed back in chapter nine, fans will support these creators as if they are family, the same way you would buy your little sibling's lemonade even if it was terrible just to show your support. The early success should be attributed to creators' massive following along with the fact that they are loved by their communities. This is what the future of this new digital world holds. More businesses like this will pop up with content creator faces on them in the years to come. If companies and brands are smart, they will look to position themselves with these unorthodox creators and see how they could leverage their resources to help these creators in exchange for a piece of the pie.

CHAPTER FOURTEEN

READY OR NOT, HERE COMES WEB3

*"At first baffling, but Web3 is growing more
mainstream and tech companies are taking note."*

—BOBBY ALLYN

Now we land on what is the most exciting chapter for
me to write, the chapter about Web3. By now many of
you may have heard of Web3 at some point in passing or
at least have heard some of the buzzwords like cryptocurrency, Bitcoin, Ethereum, NFTs, and blockchain. But
what is Web3? For starters, the term Web3 was coined in
2014 by Gavin Wood, also known for helping to develop
Ethereum.

*"At the most basic level, Web3 refers to a decentralized online
ecosystem based on the blockchain. Platforms and apps built
on Web3 won't be owned by a central gatekeeper, but rather*

by users, who will earn their ownership stake by helping to develop and maintain those services."

—GILAD EDELMAN (EDELMAN, 2021)

Before we dive deeper into what Web3 is, I'm sure many of you are wondering, if there's a Web3, then what was Web1 and Web2? I think it is important to touch on this a bit so that we truly understand the impact that Web3 is going to have. Web1 was the first phase of the internet; not much was known yet and not many people were even online at this point. Most web surfing at that time was primarily navigating to individual static webpages, the few that existed to choose from at that time. If you refer back to chapter one, Web1 occurred before any of those popular social networks were developed.

Web2 is the spot we are currently on the last legs, of here in 2022. It is the time of centralization on the internet. Most of the communication as well as the commerce is happening on platforms owned by a select few companies and is also subject to government regulation.

Web3 is an opportunity to take control out of the hands of the government and to remove some of the power from these huge corporations. That is one of the reasons cryptocurrency started to blow up, a decentralized currency that wasn't subject to a bunch of fees and that you had the power to move around instantly. The most popular of these cryptocurrencies was Bitcoin, which by now you have all heard of one way or another, whether from the twenty-year-old son of the neighbors who became a multimillionaire overnight or the skeptical people who denounce the currency every opportunity they get.

Now let's dive into a buzzword that I'm sure you've heard a lot recently: NFTs or non-fungible tokens. What does that even mean? The most simple explanation is that it's sort of like a one-of-a-kind trading card that appears as a token on a particular blockchain. A more complicated explanation is that NFTs are cryptographic assets on a blockchain with unique identification codes and metadata that distinguish them from each other. Unlike cryptocurrencies, they cannot be traded or exchanged at equivalency. This differs from fungible tokens like cryptocurrencies, which are identical to each other and, therefore, can serve as a medium for commercial transactions (Sharma, 2022). A blockchain is a distributed database. In most cases for Web3, it is used in a decentralized way to allow all users to have control.

Something else about NFTs that makes them more interesting and appealing for consumers is smart contracts. "A smart contract is programming that exists within the blockchain. This enables the network to store the information that is indicated in an NFT transaction. Once done, this information can be accessed when needed. The smart contract also ensures that the information stored is transparent as well as immutable" (AlexWGomezz, 2022).

To break that down into simpler thoughts, smart contracts are what make it possible for these NFTs to be one of a kind as well as make it possible to have permanent identification information. Smart contracts also hold the terms of agreement between the buyer and seller, written directly into the code.

When people started realizing the possibilities with NFTs, projects started popping up all over the place and sales were being made. Many individuals and companies started exploring how they could break into the space. A notable transaction that happened in the space was when Twitter cofounder Jack Dorsey sold his first ever tweet as an NFT for more than $2.9 million dollars (Conti, 2022).

Many individuals did other things similar to what Jack Dorsey did, but organizations and individuals also created entire art projects that became popular and dominated the space. I had the opportunity to talk to the creator of one of these NFT projects, Brandon Iverson, and the name of the project is Meta Moguls (Meta Moguls NFT, 2022).

Brandon is a college graduate from the McDonough Business School at Georgetown University class of 2020, and this is where I met him. We were students at the same time and even participated in some clubs on campus together which allowed us to forge a friendship. When I realized that he was venturing into the Web3 space, I knew that he would be the perfect person to talk to and follow along on this journey in the space.

One of the first things we discussed was how one of these projects even started. The first part, as you all can guess, was a simple idea. What is it that you want to share with the world? Brandon was always an entrepreneur. From a young age he worked on a multitude of different ventures, typically with his business partner and best friend Jordan Williams. It was only natural that when the NFT world started gaining some traction and he got his feet wet in the space that he would want to start his own project, and who better to do so with than his long-time business partner. Brandon and Jordan previously worked on a clothing brand together called Young Moguls Brand.

They saw the Web3 space as a more effective way of carrying out their mission to help empower creatives, entrepreneurs, and self-starters. They want to do things a little differently though. They didn't just want to create a fun art project; they wanted to build a brand. Most NFTs today don't have any real utility. A lot of them have be referred to as "cute profile picture projects," but that's not what Brandon and Jordan envisioned for the Moguls. So, they worked hard to come up with a roadmap that sought to provide it's holders with real value for being a holder.

Once they locked down their idea, they were then tasked with building out the rest of their team. They knew that they had something great on their hands, but it would require some help. They went out and recruited someone to focus on business development and content strategy. They also went out and got another team member onboard to be the community manager. Then it was time to find an artist, so they searched Fiverr, a platform that serves as a marketplace for freelancers, and they also searched on social media sites (Fiverr, 2022). It wasn't long after that they came across a talented comic book creator, Breyden Boyd, and together they came up with the out-of-the-box design.

The artwork featured these box-headed people with different things coming out of the top of the box, which represented ideas not being boxed in. They then went out and found a couple of developers to finish bringing the dream to life. In total, one hundred and seventy Meta Moguls were drawn which, in turn, led to five thousand generated.

Now it was time for arguably the most important part of the NFT project: forging a community. Could they really get people behind their idea? Would people believe in their idea as much as they did? Well, they went for it and were pleasantly surprised when they saw a community forming before their eyes. NFT projects typically start a Discord server where they build their communities. Meta Moguls had no celebrity or influencer backing. They didn't have a ton of money to throw into marketing, so seeing them amass over seven thousand people and counting in their Discord server was amazing. They did this by utilizing their own communities, as well as by tapping into the NFT communities online that are so fast with growth.

Small side note: Discord is a platform used by a ton of creators today to propel their brands. It is a place where people can communicate as well collaborate with their communities (Discord, 2022). The community can also communicate and collaborate amongst themselves already knowing they have something in common. The platform is also unique in the fact that it gives the creator so much control of their server space.

The like-minded community they built was actively communicating in the Discord basically all day about the Web3 space and how to get ahead in it. The team had some hiccups along the way with the technical aspect but quickly resolved any issues, but obstacles are to be expected in this new environment that we are just discovering. It was amazing to watch the community rally together during the hiccups and the founding team work hard to troubleshoot and fix problems. Brandon believed in the project so much that he even left his full-time job as a financial advisor to pursue the venture full time. Everyone's belief and dedication to the project made me a believer. At the time of the first launch of their NFTs, I originally planned on minting one to show support for a friend, but I later found myself with six total bought, and I have no regrets. Minting an NFT means that you are the first to purchase the token before it's sold on the secondary market. At the time of writing, they have already launched their first wave of Meta Mogul NFTs which are currently almost sold out, and by the time you are reading this, I expect them to be on their second or third wave of Moguls.

"People who are looking to start or get into the space: this era is very similar to the .com space.

Pay attention to the projects that are here for the long haul."

—BRANDON IVERSON

Another big topic in the Web3 space right now is something referred to as the metaverse. A lot of time, money, and resources are getting invested into this thing by major players in the Web3 space. The metaverse is a digital space similar to that of a video game. Imagine a digitized world, and then imagine yourself in the middle of it. I compare it to a video game but not like playing the game; it's like being in the game. There's a lot of focus going into virtual reality and augmented reality technology, to go along with the metaverse. This is how we will get the feeling of actually being in this created world. This is a very unique time, though, because we haven't reached a consensus on what will be the main platform for the metaverse which company will create the infrastructure for the metaverse that the majority of people will gravitate toward. This is a big deal, as people are already spending upwards of six figures to secure land in the metaverse. People have their assumptions, but no one can be sure, so it will also be important to keep an eye out here for new progressions.

NFTs, the metaverse, and the entire Web3 space are very new and in their infant stages. We still have so much more to learn and explore in the space, but I am a firm believer that these are going to be a part of our future in a major way. They may take a form different from what we expect today, but they will certainly be a part of our lives.

It is going to be important to keep an eye on the entire space and keep up with what moves are happening. Pay attention to the big corporations creating their own NFT projects and entering the space. Things are moving very fast, and if you don't want to be left behind you need to get aboard the train because it's not stopping for anyone. I'm not a financial advisor so I won't tell you what to invest in, but I would be remiss of me if I didn't say you should at least be watching the Web3 space. The beauty of this new world that we are living in is that in order to learn more, all you have to do is search these key terms online, read the articles and watch the videos. So much exists out there to learn here, and all that you need to know is available online.

CHAPTER FIFTEEN

BEING HEARD IN A NOISY WORLD

"Do you think of your audience first or don't you?"

—GARY VAYNERCHUK

This new digital world comes with a lot. One of the biggest things that it comes with is a noisy world. The rise of social media has led to the oversaturation of content. With people having all this access to so much content what arose was a new problem for content creators, business owners, or anyone who needed an audience. How do you get seen in this new world? I believe Gary said it best in the quote above when he talked about thinking about your audience first. So many people create content that they want without considering what their audience might want (Vaynerchuk, 2022).

After spending months researching multiple different types of content creators, the successful ones as well as the unsuccessful ones, I was able to conjure up a list of the most important things to focus on if you want to be seen in this new digital world. This is not a guarantee that you will become the next big internet personality or that your business will explode. It is more of a framework that has been followed by other successful content creators.

THE TWO ES

The key to winning online right now is by making sure your content contains at least one of the two Es: education and entertainment. Those two forms of content are thriving at the moment on the internet and social platforms. From long-form content to short-form content, the most-consumed videos are either entertaining the audience or teaching them something. This does not mean that this is the only way. Other types of content still manage to do well, but it just happens that right now entertainment and education are what are attracting the most eyeballs.

BE ORIGINAL

The next key to winning online is to be original. Everyone can identify a blatant copycat, and a copycat is typically never as good as the original. Why would the audience watch you when they could watch the person who did it better the first time? The audience will let you know in the comments that you are copying content, which is also not a good look. Once you branded as someone who steals content, it is very difficult to come back (Kane, 2022).

BE CONSISTENT

The third key to winning online is to be consistent with posting your content. All the most successful content creators post content on a consistent basis. They typically have schedules that they stick to that gives their viewers a time to look out for their content to drop. Staying relevant in the content-creation space without consistency is impossible. Too much content is available which makes it easy for people to forget all about you and the one video you posted a month ago. This does not mean that you have to post daily, but you definitely need to be posting at least once a week (Simpson, 2019).

BE FIRST

This next key to winning online is not a mandatory one, but it definitely helps to be first to something. The first person known for something is never forgotten. For example, if you are a gaming content creator and a new game comes out, being the first person to post your video on the game would prove to be very beneficial. Sometimes content creators will also get brand deals with game companies to play a game early. If the game becomes popular, you're setting yourself up for success by being one of the first to play it. This remains true for all different aspects of content. Being the first to create a video on a topic is always going to do you well. Also being the first to take over the next social platform could position you to being huge. We don't know what the next platform will be or when it's coming ,but we do know that it definitely is coming at some point. Be the first person willing to venture into that new space and experiment with the new type of content.

One truly valuable piece of advice that I can leave with content creators today is to pay attention to Web3: it's here, and it's not going anywhere. I believe that all content creators should be exploring how they can venture into the space and how they can integrate their content and brand into the space. They should be researching NFTs to see if it's for them and if creating a project makes sense for them.

FOLLOW AND TWEAK

The fifth key to winning online, follow and tweak, may seem like it is a contradiction to key number two, but it is not. For this key you want to look up some of the more successful content creators, particularly the ones who are making content in your genre. Then you should look for their most popular content, and try to analyze what about the content caused it to be successful. Then when you figure that out, take the winning concept and tweak it to make it your own. Don't copy, but instead look at what it is and make your own winning concept out of it. The best example of this is the YouTuber Airrack, whose video style seems awfully similar to some of the biggest YouTubers, yet the actual ideas of his videos are all unique to him (Airrack, 2022).

LISTEN TO YOUR AUDIENCE

The sixth and last key is to listen to your audience. Are you doing the necessary research to find out what viewers in your genre want to see? A lot of people create what they like without necessarily thinking about their audience. With so much content available to consumers it's important to figure out what people want to see. This can be done in many different ways, like looking up trending words or phrases and reading the comments on your videos as well as the comments on your peers' videos. Pay close attention to the videos that perform really well and also the ones that perform particularly bad compared to the rest. You want to try to mimic things from your high-performing videos (your audience is telling you that is what they want to see) and eliminate elements from the low-performing videos (your audience is telling you that they don't want to see that).

These points were compiled after researching multiple successful and unsuccessful content creators. I looked at why the successful ones were thriving and continued to thrive and also why the unsuccessful ones failed. Following the advice above is a sure way for a new or old content creator to succeed and thrive in this new world. The blueprint is not limited to just content creators. It can also be followed by businesses, companies, or anyone trying to grow their online presence, something that will be necessary for any type of business in the years to come. Based on our current trajectory, we will reach a point where it will be detrimental to not have an online presence. Without an online presence, it is very unlikely that you'll hold any influence in tomorrow's world.

Although this new digital world is trekking full speed ahead, you are now in a better place try to keep up.

ACKNOWLEDGMENTS

I'd like to thank the individuals that supported my book:

Amanuel Ghebremicael
Andrew (Tony) Hendrix
Anthonia Sebastien
Caitlyn Wiley
Dani Payne
Darryl Payne
Dominique Dawkins
Edward Liriano
Eric Koester
Guadalupe Avila
Joey Fernandez
Johnny Jenkins
Karthik Narayanan
Kosi Ndukwe
Madeline Moreno
Michael Abi-Habib
Naomi Dukaye
Vikash Dodani

I'd also like to thank the group of beta readers who've supported my campaign and gave feedback on my writing.

Another thank you to my editor, Chelsea Olivia, for helping me throughout the entire process. I am grateful for all her patience and knowledge along the way. Also, a big thank you to the rest of my publishing team and everyone at New Degree Press. I am forever grateful.

Special thanks go out to my family for being there and supporting me throughout the process, especially my mother. None of this would be possible without her, she has always believed in me and supported me. Whenever I needed anything, she was always there. At times, I didn't know I would make it through certain situations, but my mother always made sure I did.

Lastly, I want to thank every single reader out there. I appreciate you taking the time to read what I wrote. It means the world to me.

APPENDIX

INTRODUCTION

Barbera, J., Hanna, W., Nichols, C., Messick, D., Kasem, C., Welker, F., Christopherson, S. *Scooby-Doo, Where Are You!* Warner Home Video (Firm). Burbank, CA: Distributed by Warner Home Video. 2004.

California Cable & Telecommunications Association. "History of Cable." Learn. March 18, 2022. https://calcable.org/learn/history-of-cable/.

Geeter, Darren. "Twitch created a business around watching video games — here's how Amazon has changed the service since buying it in 2014." *CNBC News*, Updated February 26, 2019. https://www.cnbc.com/2019/02/26/history-of-twitch-gaming-livestreaming-and-YouTube.html.

G., Nick. "19 Cord Cutting Statistics and Trends in 2021 [The Dusk Of TV is Here]." *Techjury* (blog), last updated July 4th, 2021. https://techjury.net/blog/cord-cutting-statistics/#gref.

Hosch, William L.. "YouTube." Encyclopedia Britannica, December 15, 2021. https://www.britannica.com/topic/YouTube.

McFadden, Christopher. "YouTube's History and Its Impact on the Internet." *Interesting Engineering*, updated May 20, 2021. https://interestingengineering.com/YouTubes-history-and-its-impact-on-the-internet.

Muthoni, Jonas. "The Evolution Of Content And What It Means For Business Success." *Forbes*. November 16, 2020. Accessed March 17, 2022. https://www.forbes.com/sites/forbesagencycouncil/2020/11/16/the-evolution-of-content-and-what-it-means-for-business-success/?sh=7c5cae7665b0.

Nieva, Richard. "YouTube started as an online dating site." *CNET | Tech*, March 14, 2016. Accessed March 17, 2022. https://www.cnet.com/tech/services-and-software/YouTube-started-as-an-online-dating-site/.

Tidy, Joe and Sophia Smith Galer. "TikTok: The story of a social media giant." *BBC News*, August 5, 2022. https://www.bbc.com/news/technology-53640724.

CHAPTER 1

Bloomberg. "YouthStream Acquires sixdegrees in $125 Million Deal, Creating." Bloomberg press release, December 15, 1999. Bloomberg Business website. Accessed March 17, 2022. https://www.bloomberg.com/press-releases/1999-12-15/youthstream-acquires-sixdegrees-in-125-million-deal-creating.

Blystone, Dan. "The Story of Instagram: The Rise of the #1 Photo-Sharing Application." *Investopedia*, updated June 6, 2020. Accessed March 17, 2022. https://www.investopedia.com/articles/investing/102615/story-instagram-rise-1-photoosharing-app.asp.

Cook, James. "Twitch Founder: We Turned A 'Terrible Idea' Into A Billion-Dollar Company." *Business Insider*, October 20, 2014. Accessed March 17, 2022. https://www.businessinsider.com/the-story-of-video-game-streaming-site-twitch-2014-10.

Dredge, Stuart. "YouTube was meant to be a video-dating website." *The Guardian*, March 16, 2016. Accessed March 17, 2022. https://www.theguardian.com/technology/2016/mar/16/YouTube-past-video-dating-website.

Epstein, Adam. "Twitch is the undisputed champion of video game streaming." *Quartz*, updated June 10, 2021. https://qz.com/1966986/twitch-owned-by-amazon-is-the-dominant-force-in-live-streaming/.

Fiegerman, Seth. "Friendster Founder Tells His Side of the Story, 10 Years After Facebook." *Mashable*, February 3, 2014. Accessed March 17, 2022. https://mashable.com/2014/02/03/jonathan-abrams-friendster-facebook/.

Iqbal, Mansoor. "Twitch Revenue and Usage Statistics (2022)." *Business of Apps*, updated January 11, 2022. https://www.businessofapps.com/data/twitch-statistics/.

Jones, Matthew. "The Complete History of Social Media: A Timeline of the Invention of Online Networking." Technology, *History Cooperative*, June 16, 2015. Accessed March 17, 2022. https://historycooperative.org/the-history-of-social-media/.

LinkedIn. "About LinkedIn." About. Accessed March 17, 2022. https://about.linkedin.com/.

Lin, Ying. "10 Reddit Statistics Every Marketer Should Know in 2021 [Infographic]." *Oberlo* (blog), May 11, 2021. https://www.oberlo.com/blog/reddit-statistics.

Lin, Ying. "10 Twitter Statistics Every Marketer Should Know in 2021 [Infographic]." *Oberlo* (blog), January 25, 2021. https://www.oberlo.com/blog/twitter-statistics.

Maryville Unversity (blog). "The Evolution of Social Media: How Did It Begin, and Where Could It Go Next?" Accessed March 17, 2022. https://online.maryville.edu/blog/evolution-social-media/.

McMillian, Robert. "The Friendster Autopsy: How a Social Network Dies." *Wired Magazine*, February 27, 2013. Accessed March 17, 2022. https://www.wired.com/2013/02/friendster-autopsy/.

Mohsin, Maryam. "10 Instagram Statistics That You Need To Know in 2021 [Infographic]." *Oberlo* (blog), February 16, 2021. https://www.oberlo.com/blog/instagram-stats-every-marketer-should-know.

Mohsin, Maryam. "10 Tiktok Statistics That You Need To Know in 2021 [Infographic]." *Oberlo* (blog), February 16, 2021. https://www.oberlo.com/blog/tiktok-statistics.

Moreau, Elise. "Is Myspace Dead or Does It Still Exist?" *Lifewire*, updated on January 21, 2022. Accessed December 16, 2020. https://www.lifewire.com/is-myspace-dead-3486012.

Moradian, Mike. "The History of Reddit." *Honor Society*, August 17, 2020. Accessed March 18, 2022. https://www.honorsociety.org/articles/history-reddit.

Ngak, Chenda. "Then and now: a history of social networking sites." *CBS News*, July 6, 2011. Accessed March 17, 2022. https://www.cbsnews.com/pictures/then-and-now-a-history-of-social-networking-sites/.

O'Connell, Brian. "History of Snapchat: Timeline and Facts." *The Street*, February 28, 2020. Accessed March 17, 2022. https://www.thestreet.com/technology/history-of-snapchat.

Rodriguez, Salvador. "Snap reaches 500 million monthly users." *CNBC News*, May 20, 2021. https://www.cnbc.com/2021/05/20/snap-reaches-500-million-monthly-users.html.

Statista Research Department. "Number of social network users worldwide from 2017 to 2025 (in billions)." Social Media & User-Generated Content. Internet. Statista. Published January 28, 2022. https://www.statista.com/statistics/278414/number-of-worldwide-social-network-users/.

YouTube. "About YouTube." Accessed March 17, 2022. https://www.YouTube.com/intl/en-GB/about/press/.

CHAPTER 2

Chen, Tanya. "A Man Who Gained A Million Followers On TikTok Overnight Has Been Homeless And Filming In His Car." *Buzzfeed News*, September 2, 2020. https://www.buzzfeednews.com/article/tanyachen/man-behind-angry-reactions-tiktok.

D' Amelio, Oneya @angryreactions. Tiktok. Accessed March 17, 2022. https://www.tiktok.com/@angryreactions?lang=en.

Montgomery, April and Ken Mingis. "The evolution of Apple's iPhone." *Computerworld*, September 23, 2021. https://www.computerworld.com/article/2604020/the-evolution-of-apples-iphone.html.

Sungailaite, Irmante. "Homeless Teen Goes Viral With 19M Views After Showing How He Prepares His Food." Boredpanda. https://www.boredpanda.com/homeless-teen-tiktok-meals/?utm_source=google&utm_medium=organic&utm_campaign=organic.

Zeemer @randomhomelessguy2. Tiktok. Accessed March 17, 2022. https://www.tiktok.com/@randomhomelessguy2?lang=en.

CHAPTER 3

Dobrik, David and Jason Nash. "Buying My Assistant a Soccer Team." *VIEWS with David Dobrik and & Jason Nash*. December 2020. Podcast, Spotify, 35:28. https://open.spotify.com/episode/2MnBycS7a294zVvxtX8YGy?si=e2_89Z5CQKuo3X-5scpHtmg.

Gashi, Linda. Social Media Influencers—why we cannot ignore them. http://www.diva-portal.org/smash/get/diva2:1149282/FULLTEXT01.pdf.

Lexico.com. s.v. "Web 2.0." Accessed March 19, 2022. https://www.lexico.com/en/definition/web_2.0.

Rebecca Brooks. "Gary Vaynerchuk on Conan O'Brien May 2008." October 14, 2009. Accessed March 19, 2022. https://www.YouTube.com/watch?v=K1_1zqX7pls.

Reddy, Naveen. "Inspiring Biography of Gary Vaynerchuk (Wiki)." Youth Motivator, April 24, 2021. https://youth-motivator4life.com/gary-vaynerchuk-biography/.

Vaynerchuk, Gary. "Gary Vaynerchuk." My Story. Accessed March 17, 2022. https://www.garyvaynerchuk.com/biography/.

Vaynerchuk, Gary. *Jab, Jab, Jab, Right Hook: How to Tell Your Story in a Noisy World*. First Edition. New York: Harper Business, 2013.

WineLibraryTV. YouTube Channel. Accessed March 18, 2022. https://www.YouTube.com/user/WineLibraryTV.

CHAPTER 4

Aade, Mrunal. "Who Is DDG (Rapper) & What is He Famous For?" Entertainment, OtakuArt, April 17, 2021. https://otakukart.com/who-is-ddg-rapper/.

Cline, Georgette. "2021 XXL Freshman Class Revealed." XXL *Magazine*, June 16, 2021. https://www.xxlmag.com/2021-xxl-freshman-class-revealed/.

DDG. YouTube Channel. Accessed March 18, 2022. https://www.YouTube.com/channel/UCKqqDlf6lfo3ChRA4-gzusQ.

Hoorae Media, An Issa Rae Company. *The Misadventures of Awkward Black Girl (ABG)*. Accessed March 18, 2022. https://www.YouTube.com/playlist?list=PL854514FC0EBDCD8E.

Malivindi, Diandra. "11 Celebrities Who Have Spoken Out About Racial Injustice In Hollywood." Marie Claire, July 8, 2020. Accessed March 17, 2022. https://www.marieclaire.com.au/celebrities-experiences-with-racism-hollywood.

Nwandu, Angelica. "Issa Rae: 'There Was No Blueprint for My Career.'" *Glamour*, September 4, 2018. Accessed March 17, 2022. https://www.glamour.com/story/issa-rae-october-2018-cover-story.

Rae, Issa and Larry Wilmore, creators. *Insecure*. Aired on October 9, 2016, on HBO Max. https://www.hbo.com/insecure.

RDCworld1. "Anime House." January 21, 2019. Accessed March 17, 2022. https://www.YouTube.com/watch?v=_YCPJyTZ-rVw&list=PLWe2mUiIvoshIaysN2v6h_AfstRy4doLA.

RDCworld1. "Video Game House." September 14, 2016. Accessed March 17, 2022. https://www.YouTube.com/watch?v=ETY-iVLA60K0&list=PLWe2mUiIvosiHlvEWwawivoC4l7uvN-jKu.

RDCworld1 @RDCWorld1. YouTube Channel. Accessed March 17, 2022. https://www.YouTube.com/c/rdcworld1.

Smith, Will @willsmith. Instagram post. Accessed March 17, 2022. https://www.instagram.com/p/CQ9R4Zih54n/.

White, Julia @JuliaWhite. "How one talented author has secured her debut publishing deal through the power of Twitter." *Twitter* (blog), February 26, 2016. Accessed March 19, 2022. https://blog.twitter.com/en_gb/a/en-gb/2016/how-one-talented-author-has-secured-her-debut-publishing-deal-through-the-power-of-twitter.

CHAPTER 5

Airrack. YouTube channel. Accessed March 17, 2022. https://www.YouTube.com/c/airrack.

Barnhart, Brent. "Everything you need to know about social media algorithms." *Sproutsocial*, March 26, 2021. https://sproutsocial.com/insights/social-media-algorithms/.

Feeney, John. "The 10 most-subscribed YouTube channels in the world." *Prestige*, February 19, 2022. https://www.prestigeon-line.com/sg/pursuits/tech/the-most-subscribed-YouTube-channels-in-the-world/.

Leskin, Paige, Melia Russell and Steven Asarch. "Meet the 22-year-old YouTube star MrBeast, who's famous for giving away millions of dollars to strangers." *Business Insider*, updated May 5, 2021. https://www.businessinsider.com/mrbeast-YouTube-jimmy-donaldson-net-worth-life-career-challenges-teamtrees-2019-11#mrbeast-was-born-as-jimmy-donaldson-on-may-7-1998-1.

MrBeast. "How Much Money Does Pewdiepie Make????? (Updated 2015)." 2015. Video, 4:09. https://www.YouTube.com/watch?v=A8IkOqb_7Yg.

MrBeast. "I Counted To 100,000!" January 9, 2017. Video, 23:48:05. https://www.YouTube.com/watch?v=xWcldHxHFpo.

MrBeast. YouTube channel. Accessed March 17, 2022. https://www.YouTube.com/user/MrBeast6000.

Shaw, Lucas and Mark Bergen. "The North Carolina Kid Who Cracked YouTube's Secret Code." *Bloomberg News*, December 22, 2020. Accessed March 17, 2022. https://www.bloomberg.com/news/articles/2020-12-22/who-is-mrbeast-meet-You-Tube-s-top-creator-of-2020.

CHAPTER 6

Chalk, Andy. "Ninja was right: Faze Jarvis shouldn't be perm-abanned." *PC Gamer*, published November 15, 2019. Accessed March 17, 2022. https://www.pcgamer.com/ninja-was-right-faze-jarvis-shouldnt-be-permabanned/.

Epic Games. *Fortnite: Battle Royale*. Epic Games. PC/Mac. 2018.

ESportspedia. "Jarvis." Streamers. Updated July 7, 2021. Accessed March 17, 2022. https://esportspedia.com/streamers/index.php?title=Jarvis.

Jarvis. "FaZe Jarvis - Banned 4 Life (Official Music Video)." January 12, 2020. Video, 3:10. https://www.YouTube.com/watch?v=eEqFCkpaHAY.

Jarvis. "I Tricked The Internet Into Thinking I Played Fortnite." September 13, 2020. Accessed March 20, 2022. https://www.YouTube.com/watch?v=3bKwtJT769g.

Jarvis. "I've been Banned from Fortnite (I'm Sorry)." November 3, 2019. Video, 6:47. https://www.YouTube.com/watch?v=iN3t-tHug-BU.

Jarvis. YouTube Channel. Accessed March 17, 2022. https://www.YouTube.com/user/TheOnlyJaaY.

Kay. YouTube Channel. Accessed March 18, 2022. https://www.YouTube.com/kay.

Nelson, Alex. "What is an aimbot in Fortnite? Why FaZe Jarvis was banned permanently for cheating." *Inews*, November 5, 2019. Accessed March 20, 2022. https://inews.co.uk/culture/gaming/aimbot-fortnite-what-explained-faze-jarvis-banned-life-cheating-359117.

O' Connor, Florence and Zoe Haylock. "A Timeline of the David Dobrik Allegations and Controversies." *Vulture*, updated March 14, 2022. https://www.vulture.com/2022/03/david-dobrik-allegations-controversies-timeline.html.

Romano, Aja. "The sexual assault allegations against Kevin Spacey span decades. Here's what we know." *Vox*, updated December 24, 2018. Accessed March 17, 2022. https://www.vox.com/culture/2017/11/3/16602628/kevin-spacey-sexual-assault-allegations-house-of-cards.

Treyarch. *Call of Duty: Black Ops II*. Activision. PS4. 2012.

Webb, Kevin. "A teenager's lifetime ban from 'Fortnite' sheds light on a dark reality in the esports business." *Business Insider*, November 10, 2019. Accessed March 20, 2022. https://www.businessinsider.com/jarvis-fortnite-ban-epic-games-esports-ninja-2019-11#tyler-ninja-blevins-the-worlds-most-popular-fortnite-player-defends-jarvis-as-a-content-creator-1.

Willimon, Beau, creator. *House of Cards*. Aired February 1, 2013, on Netflix. https://www.netflix.com/title/70178217.

CHAPTER 7

Adin Ross @adinross. Twitter Profile. Accessed March 18, 2022. https://www.twitch.tv/adinross.

Poonia, Gitanjali. "Highest paid YouTuber MrBeast shares the formula for going viral." *Deseret News*, March 3, 2022. https://www.deseret.com/2022/3/3/22890229/highest-paid-YouTuber-mrbeast-shares-the-formula-for-going-viral.

Rockstar North. *Grand Theft Auto V*. Rockstar Games. XBox. 2013.

Skelton, Eric. "The Strange Rise of Adin Ross, Explained." *Complex*, April 29, 2021. https://www.complex.com/music/adin-ross-explained/who-is-adin-ross.

Veix, Joe. "The Strange History of One of the Internet's First Viral Videos." *Wired*, January 12, 2010. Accessed March 17, 2022. https://www.wired.com/story/history-of-the-first-viral-video/.

CHAPTER 8

Biography.com Editors. "Floyd Mayweather Biography." The Biography.com website, November 22, 2019. Accessed March 19, 2022. https://www.biography.com/athlete/floyd-mayweather.

Bohn, Mike. "Jake Paul's 'Disrupting the Space' of Boxing—But Why Is He Doing It?" *Rolling Stone*, April 7, 2021. https://www.rollingstone.com/culture/culture-features/jake-paul-boxing-career-1151995/.

Chen, Joyce. "Logan and Jake Paul: Everything You Need to Know About The YouTube Megastars." *Rolling Stone*, January 3, 2018. Accessed March 17, 2022. https://www.rollingstone.com/culture/culture-news/logan-and-jake-paul-everything-you-need-to-know-about-the-YouTube-megastars-203389/.

Dawson, Alan. "The 55 best-selling pay-per-view fight nights in history." *Business Insider*, January 24, 2020. https://www.businessinsider.com/the-50-best-selling-pay-per-view-events-boxing-ufc-wrestling-tv-history-2017-8#55-georg-es-st-pierre-v-nick-diaz-950000-ppv-buys-1.

Golliver, Ben. "The 'Bad Boys' Pistons did more than just bully Michael Jordan." *The Washington Post*, April 27, 2020. Accessed March 20, 2022. https://www.washingtonpost. com/sports/2020/04/27/bad-boys-pistons-did-more-than-just-bully-michael-jordan/.

Jake Paul. YouTube Channel. Accessed March 17, 2022. https:// www.YouTube.com/channel/UCcgVECVN4OKV6DH-1jLkqmcA.

Jake Paul. "65 Million Dollars Generated." April 18, 2021. Video, 0:39. https://www.YouTube.com/watch?v=djJQe1R404U.

Leskin, Paige and Rachel E. Greenspan. "The rise of Jake Paul, the YouTube megastar whose home was raided by the FBI as part of an ongoing investigation." *Business Insider*, updated August 6, 2020. https://www.businessinsider.com/ jake-paul-net-worth-life-career-tana-mongeau-wedding-vine-2019-10.

Mendoza, Linda, dir. *Bizaardvark*. Aired June 24, 2016, on Disney Plus. https://www.disneyplus.com/series/bizaard-vark/1yjjJRrPIYbf.

Parker, Garrett. "The 20 Richest Boxers in History." *Money Inc.* (blog), accessed March 14, 2022. https://moneyinc. com/20-richest-boxers-history/.

Rawden, Mack. "Jake Paul Knows People Hate Him, And Of Course He's Got Thoughts About That." *Cinemablend*, July 18, 2021. https://www.cinemablend.com/television/2570607/ jake-paul-knows-people-hate-him-and-of-course-hes-got-thoughts-about-that.

CHAPTER 9

Abramovitch, Seth. "Logan Paul Would Like One More Chance: "I Hate Being Hated". *Hollywood Reporter*, October 31, 2018. Accessed March 19, 2022. https://www.hollywoodreporter. com/tv/tv-features/how-YouTubes-logan-paul-reveals-plan-redeem-himself-1156187/.

Bennett, Nomi-Kate, Amy Rossmeisl, Karisma Turner, Billy D. Holcombe, Robin Young, Tiffany Brown, and Heather Key. "Parasocial Relationships: The Nature of Celebrity Fascinations." National Register of Health Service Psychologist. Accessed March 15, 2022. https://www.findapsychologist. org/parasocial-relationships-the-nature-of-celebrity-fascinations/.

Cox, Brendan. "Logan Paul's Maverick Club is Genius, here's why." *Influencive*, April 2, 2021. https://www.influencive. com/logan-pauls-maverick-club-is-genius-heres-why/.

Full Send by Nelk. "Home." Accessed March 15, 2022. https:// fullsend.com/.

MaverickClub. "Home." Accessed March 15, 2022. https://club. maverickclothing.com.

NELK BOYS @Nelk. YouTube. Accessed March 15, 2022. https://www.YouTube.com/user/NelkFilmz.

NELK. "Fake Tow Truck Driver Prank! (ARRESTED FOR GTA)." June 22, 2021. Video, 24:55. https://www.YouTube.com/watch?v=xsbD4MB0OT8.

Sam Saffa. "NELK | How Much They Make REVEALED | The Reality of Full Send." June 1, 2021. Video, 10:38. https://www.YouTube.com/watch?v=C4AijvM72-s.

Seek Discomfort. "Home." Accessed March 15, 2022. https://www.seekdiscomfort.com/

CHAPTER 10

Flamingo. YouTube Channel. Accessed March 18, 2022. https://www.YouTube.com/c/flamingo.

Griffith, Eric. "Most Parents Say Games Have a Positive Impact on Their Kids." *PCMag*, August 23, 2021. https://www.pcmag.com/news/most-parents-say-games-have-a-positive-impact-on-their-kids.

RowdyRogan @RowdyRogan. Twitter account page. Accessed March 17, 2022. https://twitter.com/RowdyRogan?ref_src=twsrc%5Egoogle%7Ctwcamp%5Eserp%7Ctwgr%5Eauthor.

VIEWS. "Charli and Dixie on Being Famous Sisters." February 24, 2021. Video, 45:05. https://www.YouTube.com/watch?v=9Y-Q19x66U6E.

CHAPTER 11

Bell, Karissa. "Instagram is working on creator shops and a 'branded content marketplace' for influencers." *Engadget*, April 27, 2021. https://www.engadget.com/instagram-creator-shops-content-marketplace-190211314.html.

Boomer, Jim. "Change Management in Accounting Firms." *Firm Management | CPA Practive Advisor*, September 8, 2016. Accessed March 15, 2022. https://www.cpapracticeadvisor.com/firm-management/article/12248264/change-management-in-accounting-firms.

Demeku, Amanda. "Instagram Announces New Features for Creators to Make Money." *Laterblog*, June 8, 2021. https://later.com/blog/instagram-creator-monetization-features/.

Johnson, Joseph. "Global digital population as of January 2021." Demographics & use. Statista. September 10, 2021. https://www.statista.com/statistics/617136/digital-population-worldwide/.

Tapp, Tom. "Disney Plus: Everything You Need to Know About Disney's Subscription Streaming Service." *Next TV*, February 9, 2022. https://www.nexttv.com/news/disney-plus.

CHAPTER 12

Faze Clan. "Home." Accessed March 19, 2022. https://fazeclan.com/.

Forbes. "Jeremy Wang, 29." Profiles | 30Under30: Games. Accessed March 15, 2022. https://www.forbes.com/30-under-30/2021/games/?profile=jeremy-wang.

Gonzalez, Sandra. "Tom Hanks and Rita Wilson diagnosed with coronavirus." *CNN*, updated March 12, 2020. Accessed March 15, 2022. https://www.cnn.com/2020/03/11/entertainment/tom-hanks-rita-wilson-coronavirus/index.html.

Moore, Logan. "Among Us Streamer Disguised Toast Officially Waves Goodbye to the Game." Comicbook, May 1, 2021. https://comicbook.com/gaming/news/among-us-disguised-toast-twitch-YouTube-goodbye/.

Raven Software and Infinity Ward. Call of Duty: Warzone. Activision. Xbox. 2020.

Stedman, Alex. "Kris 'FaZe Swagg' Lamberson on Encouraging More Black Gamers to Join the Scene." *Variety*, February 24, 2021. https://variety.com/2021/digital/news/faze-swagg-twitch-black-gamers-streamers-1234913255/.

Stream Hatchet. COVID-19 Impact on Streaming Audiences. Barcelona, Spain: Stream Hatchet SLU, ESports Data and Analytics Company, 2020. https://strivesponsorship.com/wp-content/uploads/2020/05/COVID-19-Impact-on-Streaming-audiences-report.pdf.

Wamsley, Laurel. "March 11, 2020: The Day Everything Changed." *NPR WNYC*, March 11, 2021. https://www.npr.org/2021/03/11/975663437/march-11-2020-the-day-everything-changed.

Wong, Amy. "TikTok blew up in the U.S. in 2020; meet 6 of the hottest creators from the Seattle area." *Seattle Times*, updated December 31, 2020. Accessed March 15, 2022. https://www.seattletimes.com/life/tiktok-blew-up-in-the-u-s-in-2020-meet-6-of-the-hottest-creators-from-the-seattle-area/?amp=1.

CHAPTER 13

Alford, Aaron. "Mr. Beast's IRL Squid Game draws more views in first week than Squid Game did in a month." INVEN-Global, December 2, 2021. https://www.invenglobal.com/articles/15859/mr-beasts-irl-squid-game-draws-more-views-in-a-week-than-squid-game-did-in-a-month.

Dong-Hyuk, Hwang, dir. *Squid Games*. Season 1. Aired September 17, 2021, on Netflix. https://www.netflix.com/title/81040344.

Friedman, Vanessa and Jessica Testa. "The Metaphysics of Kylie Cosmetics Being Sold to Coty." *The New York Times*, updated October 28, 2020. Accessed March 17, 2022. https://www.nytimes.com/2019/11/19/style/kylie-jenner-coty-cosmetics.html.

Happy Dad. "Home." 2022. Accessed March 17, 2022. https://happydad.com/.

Kylie Cosmetics. "Home." Accessed March 19, 2022. https://kyliecosmetics.com/en-us.

Maze, Jonathan. "Why MrBeast Burger is the Most Important Restaurant Concept in the U.S. Right Now." Restaurant Business, February 12, 2021. https://www.restaurantbusinessonline.com/financing/why-mrbeast-burger-most-important-restaurant-concept-us-right-now.

MrBeast Burger. "About us." 2022. Accessed March 17, 2022. https://mrbeastburger.com/about.

MrBeast. YouTube channel. Accessed March 11, 2022. https://www.YouTube.com/c/MrBeast6000.

NELK. "We Made Our Own NELK Alcohol! (HAPPY DAD HARD SELTZER)." June 1, 2021. Video, 33:14. https://www.YouTube.com/watch?v=BlZ-Wc3cMP8.

Watson, Amie. "The Untold Truth Of MrBeast Burger." *Mashed*, updated January 11, 2021. https://www.mashed.com/308763/the-untold-truth-of-mrbeast-burger/.

CHAPTER 14

AlexWGomezz. "NFT Smart Contracts Explained." NFT. Cyberscrilla. Accessed March 17, 2022. https://cyberscrilla.com/nft-smart-contracts-explained/.

Conti, Robyn. "What Is An NFT? Non-Fungible Tokens Explained." Forbes Advisor, updated on February 15, 2022. https://www.forbes.com/advisor/investing/nft-non-fungible-token/.

Edelman, Gilad. "The Father of Web3 Wants You to Trust Less." *Wired*, November 29, 2021. https://www.wired.com/story/web3-gavin-wood-interview/.

Fiverr. "Home." Accessed March 18, 2022. https://www.fiverr.com/.

Meta Moguls NFT. "About." Accessed March 18, 2022. https://www.metamoguls.art/.

Sharma, Rakesh. "Non-Fungible Token (NFT) Definition." Investopedia, updated February 26, 2022. https://www.investopedia.com/non-fungible-tokens-nft-5115211.

CHAPTER 15

Airrack. YouTube channel. Accessed March 17, 2022. https://www.YouTube.com/c/airrack.

Kane. Tiktok account. Accessed March 20, 2022. https://vm.tiktok.com/ZTdPS4H2V/.

Simpson, Jon. "Why Content Consistency Is Key To Your Marketing Strategy." Forbes. February 11, 2019. Accessed March 11, 2019. https://www.forbes.com/sites/forbesagencycouncil/2019/02/11/why-content-consistency-is-key-to-your-marketing-strategy/?sh=f41a4574ef57.

Vaynerchuk, Gary. "Turn Your Company Into a Content Empire by Using These 4 Steps." *Gary Vaynerchuk* (blog). Accessed March 19, 2022. https://www.garyvaynerchuk.com/turn-your-company-into-a-content-empire-by-using-these-4-steps/.

Printed in Great Britain
by Amazon

82865186R00098